◀JOINERY▶

JOINERY

METHODS OF FASTENING WOOD

CHARLES SELF

A STOREY PUBLISHING BOOK

Storey Communications, Inc.
Pownal, Vermont 05261

Cover design by Carol Jessop

Text design and production by Michelle Arabia

Photographs and illustrations courtesy of the author with permission from manufacturers and distributors

Edited by Roger Griffith

The information in this book is true and complete to the best of our knowledge. All recommendations are made without guarantee on the part of the author or Storey Communications, Inc. The author and publisher disclaim any liability with the use of this information.

Copyright © 1991 by Storey Communications, Inc.

Printed in the United States by R.R. Donnelley
First Printing, March 1991

Library of Congress Cataloging-in-Publication Data

Self, Charles R.
    Joinery : methods of fastening wood / Charles Self.
        p.   cm.
    "A Storey Publishing book."
    Includes index.
    ISBN 0-88266-641-X (hc) — ISBN 0-88266-648-7 (pb)
    1. Joinery.   I. Title.
TH5662.S39   1991   90-50415
694'.6—dc20   CIP

CONTENTS

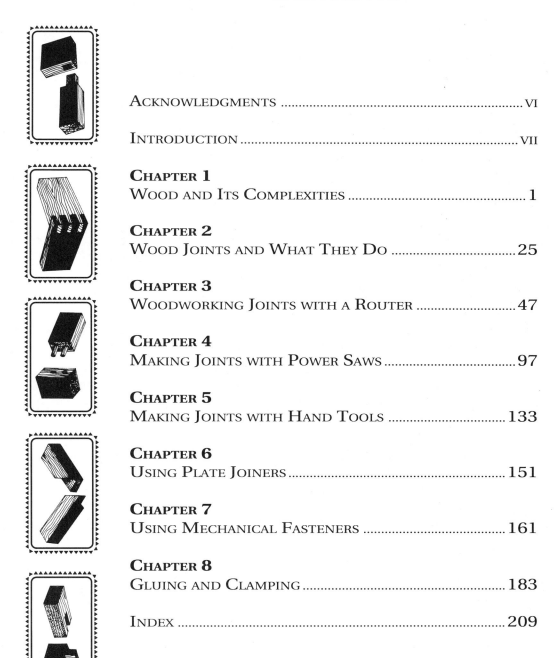

ACKNOWLEDGMENTS

Thanks must go to many people. The following were of great assistance in gathering materials, information, and illustrations.

Stan Black of Trend-Lines helped a great deal, and offers a special deal for readers of this book: a Trend-Lines catalog, usually $2, will be free to readers of this book, if you mention the book in your request for the catalog, by writing to Trend-Lines, 375 Beacham Street, Chelsea, Massachusetts 02150, or calling 1-800-767-9999.

Also David Draves of Woodcraft, Paul Thoms of the Woodworker's Store, Kim Park of Porter-Cable, Tim Miller of Vermont-American, David Springstroh of Disston, Richard Ziff of Cooper Tools, Jim Ippolito of Ryobi, Bobby Weaver of Bedford, Virginia, and many of the staff at Stanley Tools and Makita Tools.

INTRODUCTION

here are hundreds of methods of wood fastening, each meant for a different job, and many for decorative purposes only. No one person knows them all. As soon as anyone thinks he does, someone comes up with a new way, be it a mechanical or cut joint or a new or different way to cut an old style.

My purpose in writing this book is to develop and explain the methods I've learned. Learning has taken a period of years, long enough to help anyone, but I still haven't reached the pinnacle of master craftsmanship that only a few attain.

Different wood fastening methods are required for carpentry, for simple bookshelves, and for more complex furniture pieces, drawers for example. What is good for one job, such as tight-fitting indoor furniture joints, doesn't work well in other applications, such as outdoor furniture. We'll discuss many of these.

In some cases, hand tools and hand-cut joints are still used, either for aesthetic reasons or because it's less work than with power tools. But I am no purist who believes work needs to be done as my grandfather did it to be authentic. One of my grandfathers owned and operated a sawmill and a farm. He used every power option available, from mules to tractor engines to Model T-powered saws. The cost of handwork is gnarled fingers and stooped shoulders, and it is only appreciated by those who have later, easier options.

Dovetails can be made beautifully with machines and templates, with far greater ease and with great precision. Too, they don't require the brain-spraining effort of laying out the individual pins and tails properly that hand cutting demands.

Modern tools can produce accurate butt joints more easily than age-old hand-planing techniques, and don't require a half-dozen years of apprenticeship. The investment need not be a huge one: two good hand planes equal the price of the power system. Today top-quality hand tools will not save much money over good quality portable power tools. This is one point on which I am adamant: go with the best tools you can afford. You'll find your work goes faster and is more accurate, and your time spent is more enjoyable.

Most joints, as you will discover, can be cut with several different tool combinations. The dovetail is a good example. It can be cut by hand, with a router and jig setup, or on a table saw. Finger joints can be made on a table saw or with a router, but are seldom made with hand tools (Figure I-1). Mortise-and-tenon joints offer even wider variations in tool use.

Joints range from the very simple, such as the nailed or glued butt joint, on through to the exceptionally, and usu-

Fig. I-1 Finger joint in cherry, cut on a table saw. Here the fingers are larger than they would normally be, cut simply as a demonstration.

ally unnecessarily, complex. Originally a joint design was created to do a particular job well. Dovetails, for example, provide exceptional strength in drawers, as they resist forces from all the directions from which pull will be applied. Today, dovetails are used almost as frequently as decorative joints as for strength.

To achieve skill, you must spend time working on projects. You can read, imagine, and make drawings all you want, but you will never develop skill in joining wood until you start measuring, marking, cutting, and joining. Practice time is an essential ingredient of successful joint making, even in driving nails.

I prefer joints that are simple and useful, with beauty coming from a form as it fills a need, not from a form that does nothing beyond adding unneeded complexity.

Throughout this book, you'll find both factory- and shop-made jigs and aids for assistance in doing particular jobs. Once accuracy is gained, you must strive for the ability to do the same thing twice or more, and that is easier through use of jigs, such as a simple stop block to assure that all parts are the same length, or a more complex setup for producing finger or other joints, rapidly and with perfection, so that the final joint will slip together with just enough room left for glue.

Of course, joints aren't the only things used to join wood, and we'll look at the others, the many varieties of nails, screws, and bolts.

We'll also take a look at adhesives. The best woodworking adhesive has become harder to select because today there are more options. The only selections once were hide glues and some moisture-proof glues for exterior work. Today, hide glue is used infrequently by home shop workers because it requires a great deal of extra work. Many of its purposes are served, and served well, by white and yellow resin glues, while the acceptance of waterproof glues has gone from urea resins to resorcinols to epoxies, and some instant glues are now being formulated in ways that work reasonably well with woods.

Read the book quickly. Skim to those parts of most

interest to you, read those thoroughly, then place the book beside or on your workbench and start producing joints for your projects, or for one of the projects I've included to provide demonstrations of the utility of various forms of joints.

I hope you enjoy working with the book as much as I have enjoyed working on it.

WOODWORKING SAFETY

Safety when working with wood is absolutely necessary. I take care to remove my watch and jewelry before beginning, and I keep my clothes from flapping and catching in tools.

It is in the techniques, though, even after all basic safety rules are met, that safety problems may be encountered. Woodworking is inherently dangerous, a business and hobby that requires constant attention to safety to ensure the maintenance of one's full complement of digits and other appendages.

In this book I have tried to make all techniques and jigs as safe as possible, when used with reasonable care. If at any point you feel uncomfortable with a technique or a design, don't use it.

What is safe in my shop may not be safe in yours, and vice versa. It is you who must make the decision as to safety, for no matter how something is described, it is impossible for me to envision and describe all the various ways a method or device may be changed to fit circumstances.

Keep it safe for you.

CHAPTER 1

WOOD AND ITS COMPLEXITIES

odern techniques for joining wood make the work easier, but can't remove all the difficulties that wood, by its nature, challenges us with as we try to form it into useful shapes.

Wood is almost an ideal natural material. It can be shaped with tools, bent with or without steaming, and is durable under abrasion and different stresses. Some woods withstand weathering conditions that will ruin almost anything, including steel. Wood is relatively cheap and naturally attractive.

However, it does present a variety of problems to the joint maker because of one of its inherent qualities: wood moves, expanding and contracting, in response to changing humidity, even indoors.

THE NATURE OF TREES

Trees are perennial plants capable of adding annual growth. The trunk supplies most of our saw lumber, but limbs, and even roots, can provide some amazingly beautiful wood patterns. Most of us seldom see the bark on lumber, as that is stripped off before the log is sawed. The inner and outer barks cover the cambium, which surrounds the wood of the tree. This wood is divided into the inner heartwood and the outer sapwood. In the center is the pith. The typical wood cell consists of a cell wall surrounding an inner cell cavity. The cell is long, but the width-to-length ratio varies a lot by species. Most cells are

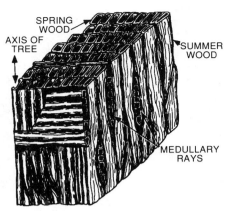

SPRING WOOD

AXIS OF TREE

SUMMER WOOD

MEDULLARY RAYS

Fig. 1-1 The structure of wood.

arranged vertically, up the structure of the trunk and limbs (Figure 1-1). These produce the grain direction, while their arrangement around each other produces the grain pattern of the wood. Woods such as southern pine, with strong contrast between early and late wood growth, have uneven grain patterns (strong differences in color and texture), while woods such as white pines show little difference in the grain color and texture.

Rays radiate from the center of a tree. They carry food laterally.

There are characteristic differences between heartwood and sapwood. Heartwood is sapwood that is no longer needed to transmit sap to the leaves, so the cells are transformed, a change that is accompanied by the formation of cell wall extractives in the heartwood. It is the heartwood of all species that is of most interest to the woodworker, because sapwood varies little in color from species to species, with most sapwoods ranging from an off-white to a light tan, and with a few shading to the yellow. Heartwoods range from the dark, lush browns of walnut, the light, lovely reds of cherry, to the whiteness of birch and some pines.

Heartwood, given pigmentation by the extractives, substances deposited in wood during its transition from sapwood to heartwood, may get other characteristics from these substances. These include resistance to decay because some extractives, such as cedars, are resistant to fungi. Others make the wood less permeable, thus making the wood dry more slowly, and sometimes making it more dense. Heartwood shrinks less than sapwood. When a tree is cut down, the sapwood may contain five times as much moisture as heartwood, which means sapwood tends to warp more in drying.

Cellular shape changes do not occur as sapwood becomes heartwood in a maturing tree, nor are cells added or subtracted. The overall shape and strength of the wood do not change.

Sapwood proportions vary greatly from tree species to tree species. Some trees have only a single annual growth ring as sapwood, while others have 80 or 100. Most popular woods have sapwood regions ranging from an 1½" to 6" wide (Figure 1-2).

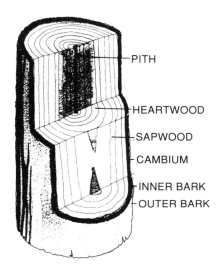

PITH

HEARTWOOD

SAPWOOD

CAMBIUM

INNER BARK

OUTER BARK

Fig. 1-2 Cross section of a tree.

Wood Density

Wood density is the weight per unit volume, usually expressed as weight per cubic foot. Density is usually measured and expressed as specific gravity, which is the ratio of the volume of the wood, in this instance, to the volume of water. The density of wood changes as the wood expands and shrinks. Because of this property, it is usual for measurements to be taken with wood dried to a specific water content. Wood density is important because denser wood shrinks or swells more. Density is also something of a predictor of ease in machining: denser (or harder) woods usually are more difficult to machine. They are also far harder to nail, but hold nails much better than less dense woods.

Hardwoods and Softwoods

Classification of woods into hardwood and softwood types is not accurate since it really involves the division between conifers (with seed-bearing cones) and deciduous trees (leaf dropping). The wood of some deciduous trees is soft, such as basswood, while the wood of some conifers, such as southern yellow pine, is hard.

The specific gravity of softwoods varies from about 0.28 to 0.55, while domestic hardwoods range from about 0.35 (cottonwood, basswood, and butternut) to 0.75 (shagbark hickory, our hardest and most dense wood). Exotics range from balsa's 0.15 up to lignum vitae's 1.23 (it does sink in water). *Specific gravity* is the ratio of a cubic foot of water displaced by a cubic foot of a solid object, with the water at 4° C.

Softwoods have a slightly less complex structure than do hardwoods. The softwood's cells are of a fiberlike construction about a hundred times longer than their width. An average softwood will contain as many as four million of these cells in a single cubic inch. Coarse-textured woods have the largest diameter cells, while fine-textured woods have smaller diameter cells; the smaller the diameter, the smoother the finish and, generally, the better the overall finishing qualities. Redwood is a superb wood for many uses, but is the coarsest textured of all softwoods, so is seldom used where a truly fine finish is required. The cells of hardwoods are thick and have small cavities; the cells of softwoods are just the opposite, thin-walled with large cavities.

Some softwoods also have resin canals or gum ducts where, in sapwood, fluid resins exist. Over time, the resins will solidify. If they haven't solidified, they will bleed to the surface after you machine the wood. Kiln drying tends to solidify the resins so that any resulting drops can be scraped or sanded off, and bleed-through does not occur. If bleed-through takes place, you'll find tiny yellow specks on the surface. The specks will show through paint.

Hardwood structure is quite different. Only a few tropical hardwoods have resin canals. Rays vary in size, but in most cases are larger than those found in softwoods. Hardwoods generally have more specialized cells, including vessel elements with very thin walls and open ends. These form up end to end and are ideal transporters of sap. Interspersed among elements will be fibers with closed ends and thick walls. Instead of conducting sap, these provide the tree with strength. The vessels are what cause hardwoods to be classed as porous woods. Vessels and fibers are distributed in different manners in different species, with oak, elm, ash, and some others having most of the vessels, or pores, in the earlywood (first growth part of the growth ring), thus creating an uneven grain structure known as ring-porous. This pattern also makes staining an uneven process. The pores of hardwoods such as maple and cherry are distributed fairly evenly, and are classed as diffuse-porous.

Some woods have no distinct zones. Larger pores in earlywood taper to smaller sizes in latewood (the part of growth ring formed after earlywood), with no dividing ring. These woods, such as black walnut and butternut (often called white walnut), are thus classed as semi-ring porous *or* semi-diffuse porous.

Another structural feature found only in hardwoods is the tylosis, a bubblelike structure that forms in the cavities of the vessel elements. Tylosis limits liquid passage. White oak is far better than red oak for making barrels and other items that store liquids because it is packed with tyloses, while red oak has few or none.

THE PATTERN OF RAYS

The rays found in hardwoods vary much more in size than do those in softwoods. Some hardwood rays are only a single cell wide and require a microscope to view, while others may be as much as forty cells wide. Single rays four inches long have been found in white oak. Rays are important to woodworkers for a number of reasons. One is that drying will cause checks in the plane of the rays. These natural planes of cleavage are handy for riving shingles and roughing out parts, but make it difficult to get a smooth surface.

Sycamore, oak, and beech have ray surfaces with attractive patterns. Ray patterns may be darker than the background, such as with maple and sycamore, or lighter, as with cherry and yellow poplar.

Sawing

The two most common methods of milling any tree are plainsawing and quartersawing through the board. Both produce primarily flat grain boards. Plainsawing, or sawing directly through the log as it is fed into the sawmill, with no changes in position, tends to produce a mix of figures, or grain patterns (Figure 1-3).

Fig. 1-3 As shown, the slash-cut method of cutting lumber is faster and cheaper because it produces more boards with less waste. Rift cutting is best for lumber to be used for cabinetry because it is more stable dimensionally. Rift cutting is often called quarter-sawing.

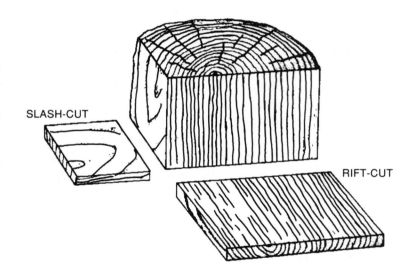

Plainsawed lumber is not the best for furniture making or for any kind of precise joint making because it mixes sapwood and heartwood. The different densities and drying needs of the sapwood and heartwood insure that the board will cup, which is to warp across the width of the board. The wider the board, the worse the cupping will be. Plainsawed lumber can be easily identified by the U- or V-shaped markings at the ends of the boards, caused by the growth ring arcs.

Quartersawed lumber is more expensive than plainsawed. It is cut basically at a 90° angle to the growth

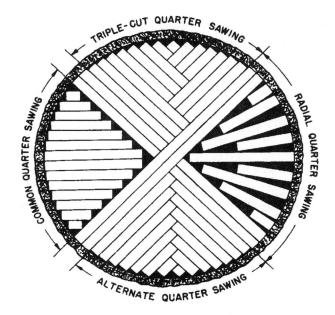

Fig. 1-4 *Four methods of quartersawing.*

rings, and is less prone to warping both before and after machining. There is less of a mix of heartwood and sapwood in each board. The log is first cut lengthwise into quarters. The quarters are then cut from the bark to the center, producing a board that will warp less and expose a more attractive grain (Figure 1-4).

Rotary cutting, or peeling, of logs produces a veneer that is used for basic plywoods. More costly veneers are produced by half-round slicing or quarter slicing. The latter uses a knife taking a very thin — often as little as $1/28$" thick — slice off a quarter of the wood, as in quartersawing. Slicing produces less waste.

Several milling methods are used to get highly patterned boards from burls, crotches, and other special woods.

WOOD AND WATER

Once you've learned to form joints accurately, most of the problems you'll face involve water in wood. All of us know the problems, from doors that are too loose in winter and too tight in summer, to cracks appearing in deck materials and outdoor furniture.

A log's weight may be 50 percent or more water when first cut. Free water in the cells evaporates first, reducing the moisture content to about 30 percent. At this point, the cells have not changed in shape, so there has been no shrinkage of the wood. When moisture content is reduced below 30 percent, water within the cell walls is reduced, and the wood may shrink and warp.

Two methods are used to dry lumber. Air drying will reduce the moisture content to between 12 and 19 percent. Kiln drying will reduce lumber's moisture content to 6 to 8 percent. That is good for lumber to be used in cabinet or furniture making.

Wood movement — shrinking and swelling — is related to the overall size of the pieces used in construction. If a project is built of large, solid wood panels, movement will be far greater than if the individual pieces are small. This is one of the reasons why plywood keeps its popularity in cabinetmaking. With its thin cross-laminated plies, plywood tends to stabilize itself, as one ply exerts force in one direction, the next exerts force in an opposing direction. Even properly glued large solid-wood panels move enough over a period of years to cause warping.

Taking natural wood movement into account is essential to accurate and durable joint making. Thus, it is important to finish wood pieces in a similar manner, using the same type of finish and thickness of finish on all sides to stabilize the wood, so one side doesn't move more than the other in reaction to different rates of moisture penetration and loss.

Wood is flexible. It is a hygroscopic material, losing and gaining water as the relative humidity of the surrounding air changes. Wood's stability depends in large part on the reduction of water in the wood, then on keeping that water out of the wood. Without finish on all sides of a project, moisture penetration and loss is unequal, causing uneven wood swelling and shrinking. The result can be warping, then destroyed joints.

To achieve stability means wood taken from forests

must be dried to about the point where the moisture content is similar to that of its environment. Environmental moisture content is a changing specification, as the humidity levels in each area vary drastically. A swing from 10 percent to 15 percent doesn't seem like much, only five percentage points, but remember it is a 50 percent change in water content of the air. We find some of the widest humidity variables in homes. In general, winter humidity levels in most homes are low, while summer levels are high. This can vary widely, even within the particular season. Hours of cooking during the winter, for example, will raise humidity levels close to those of midsummer.

A moisture meter will show a piece of wood at a 10 percent moisture level in winter, then at 20 percent in summer in the same location. That means the wood has absorbed 100 percent more moisture, thus swelling its cell walls. Conversely, in winter, the wood will lose half its moisture content, allowing cell walls to shrink. The likely wood movement is obvious.

For basic furniture making, air-dried lumber is inadequate because it contains too much moisture. For cabinetmakers, kiln drying to a moisture content of below 10 percent is necessary.

Storage of Lumber

Once dried, lumber does not stay dry. Outdoor storage of lumber kiln-dried to cabinetmaker's standards is not recommended in any but desert areas, as the lumber absorbs water in such storage, often rising to close to 15 percent in just three seasons.

Damp indoor storage is almost as bad. My shop is in a rented basement some ten miles from my home, and thirty miles from my nearest hardwood supplier. The basement walls are brick, probably over 100 years old, and tend to leak from time to time. Immediate vacuuming up of pooled water, when possible, and the use of a dehumidifier help, but not really enough to prevent some swelling in stored lumber. Thus, I store very little wood.

Dimensional Changes

Wood shrinks as moisture is removed, and swells as it is regained, as the previous pages have described. The water gained is classed as bound water — that is, water held in the cell walls.

The formulas for linear, radial, and tangential shrinkage are fairly difficult to use, and they are unimportant to most of us.

In general, total shrinkage along the grain (linear) of wood — the length of the board — is minor, about 0.02 percent. Thus, something like 0.096 of an inch is lost in every eight feet of length in normal drying.

It is the shrinkage across the wood's width, the tangential and radial shrinkage, that concerns us as woodworkers. Tangential shrinkage, the shrinkage away from the center of the log — that is, where the center would be if it still existed — is the greatest, while radial shrinkage — edge to edge on a board, effectively — is next. And these two combine to form tangential-radial shrinkage, a ratio of great importance to us.

Teak has about the lowest overall ratio of shrinkage of any wood, coming in at a combined ratio of 1:8, with a total for the two of about 6¼ percent. Among other popular woods in the workshop, we find sugar maple, with a combined shrinkage of over 12 percent, and pecan hickory over 13 percent. White oak is over 16 percent, while sugar pine is about 8½ percent. These figures are for plainsawed boards. Quartersawed boards will react differently and less severely.

Overall, we are fairly safe if we figure on a 12 percent shrinkage rate for wood, but the chart indicates that some woods are safer than others, more stable, and less apt to pull apart carefully designed and worked joints.

AMOUNT OF WOOD SHRINKAGE
FROM GREEN TO OVEN-DRY

(10 percent moisture content)

SPECIES	(TANGENTIAL AND RADIAL) COMBINED SHRINKAGE	SPECIES	(TANGENTIAL AND RADIAL) COMBINED SHRINKAGE
▶ **HARDWOODS**		CEDAR, WESTERN RED	7½
ASH, WHITE	13	DOUGLAS FIR	12¾
ASPEN	10¼	FIR	10
BASSWOOD	16	HEMLOCK, EASTERN	9¾
BEECH	17½	HEMLOCK, WESTERN	12
BIRCH, YELLOW	16½	LARCH	13½
BUTTERNUT	10	PINE, EASTERN WHITE	12¼
CHERRY, BLACK	10¾	PINE, RED	11
ELM, AMERICAN	13¾	PINE, SUGAR	8½
ELM, ROCK	13	PINE, WESTERN WHITE	11½
HICKORY, PECAN	14	REDWOOD	7
HICKORY, SHAGBARK	17½	SPRUCE	11½
MAPLE, HARD (SUGAR)	14¾		
MAPLE, RED	12	▶ **EXOTICS**	
OAK, RED	13	AVODIRE	10¼
OAK, WHITE	16	BALSA	10½
SWEET GUM	15½	GREENHEART	17¼
SYCAMORE	13½	LAUAN	11¾
WALNUT, BLACK	13¼	MAHOGANY	8¾
YELLOW POPLAR	12¾	OBECHE	8½
		PRIMAVERA	8¼
▶ **SOFTWOODS**		RAMIN	12½
CEDAR, EASTERN RED	7¾	TEAK	6¼
CEDAR, NORTHERN WHITE	7	WALNUT, EUROPEAN	10¾

(EXPRESSED IN %)

As you will note, most of our furniture woods do a fair amount of moving. Wood joints are generally loose enough to absorb this movement, no matter how tight.

Shrinkage can be partially moderated by finishes, not by metal anchors. Wood movement may also be moderated by proper selection and handling of materials.

Keeping Wood Dry

For most of us, selecting properly kiln-dried wood is a procedure that involves trust. Kiln drying is not a particularly high-tech process, but it is touchy and requires careful handling and scheduling of materials. Monitoring of selected boards is required. The equipment involved is not cheap, and even small versions tend to be well beyond the needs of home workshops.

We easily see the advantage of owning a small planer when custom planing now approaches 20 percent of the cost of a furniture quality board. The cost and care required with kiln drying tend to be high enough to deter almost all of us.

For us, storing materials in their proper environment, keeping dry what is already dry, and not storing so much we cannot keep its condition checked for proper humidity levels, are the sensible way.

For the beginning woodworker, the storage of large amounts lumber is expensive and unnecessary, but it is helpful to have enough stored for several projects. Picking up bits and pieces as they are needed will delay work and may be costly. Too often the woods needed are not available, forcing a wait while the wood is shipped — if the species is currently on the market.

My favorite wood is cherry, so I try to keep a few hundred board feet on hand, without getting fancy or excessive about storage. Wood for immediate use is set on stickers, elevated from the floor by several pressure-treated 2x4s, because of my wall leakage problem. I also keep some walnut, some pine, and a fair amount of white oak as well. Poplar is kept for drawer sides, and poplar plywood for drawer bottoms. General plywood is on hand for fast shelving, small workshop stands, tables, and similar uses. I don't cover wood, except when transporting it. I've also made some pressure-treated cross boards for my pickup to make sure, should I get caught in a heavy rain and the truck bed drains stop up, that the wood does not rest in water. On longer boards, I seal ends with paraffin; canning wax does as well as anything. End coating is not necessary unless the wood is to be stored for some time.

For stickers, I use 1x material, placed on each layer of boards, 90° to the grain, at 2' intervals along the boards.

When dealing with expensive materials, I take simple, basic precautions. Lumberyard bills will shock you when you figure the costs of some projects using hardwoods. A small stack of walnut, cherry, or oak adds up to several thousand dollars these days, so letting it warp or rot is not wise.

I recommend a good quality dehumidifier for those who have basement shops, even shops that do not have leaky walls.

Wood Selection

Softwoods generally fall under structural lumber grading methods, divided into three primary categories: appearance lumber, stress-graded lumber, and non-stress-graded lumber.

Appearance softwood grades are the highest, as they are in hardwoods, and are normally listed as Select or Finish. You will find select and finish interior grades marked B and Better (B&BTR): this is the highest grade generally available today. Appearance grades, no matter how pretty, are not necessarily the strongest, because there are no limits on such faults as grain deviations, density, growth rings per inch, and others. For structural lumber, stress-graded materials are needed.

Stress-graded materials, such as structural 2x4s, are

graded Construction, Standard, and Utility. Structural planking (2x6s, 2x8s, etc.) is graded as Select Structural No. 1, No. 2, and No. 3. Such lumber is grade-stamped at the mill, unless you are dealing with small local mills that provide you with air-dried or green unplaned lumber. The use of such lumber, in recent years, has been severely limited for residential structures. If you have any intention of using it, check building codes first or you may end up tearing down all or part of a structure.

Most softwood will be stamped KD if kiln-dried.

Hardwood lumber is sawed, in large part, as factory lumber. Supplied in random widths and lengths (though your local supplier will usually provide widths and lengths to suit), factory lumber comes in a number of grades.

The grades are Firsts, Seconds, Firsts and Seconds or FAS, Selects, and then Common, starting with No. 1 and dropping to 3B. For our purposes, anything under a No. 2 Common (minimum clear width, one face, is 3" x 2') is useless.

Most hardwood sold today is FAS. You get clear lengths no less than 3" x 7', or 4" x 5', with over 81 percent of the face clear of defects, on a board width of 6" and a length of from 8' to 16'. Dropping down to selects gives 83 percent clear of defects, but on a board width of 4" and lengths from 6' to 16'. Clear cutting requirements are the same.

Useful Woods

WOOD	LOCALE	CHARACTERISTICS
ASH	EAST OF ROCKIES	STRONG, HEAVY, TOUGH GRAIN THAT IS STRAIGHT. SOMETIMES SUBSTITUTES FOR MORE COSTLY OAKS.
BASSWOOD	EASTERN HALF OF U.S.	SOFT, LIGHT, WEAK WOOD THAT SHRINKS CONSIDERABLY, VERY UNIFORM, WORKS EASILY, DOES NOT TWIST OR WARP.
BEECH	EAST OF MISSISSIPPI, SOUTH-EASTERN CANADA	SIMILAR TO BIRCH, SHRINKS, CHECKS CONSIDERABLY, CLOSE GRAIN, MAY BE A LIGHT OR DARK RED COLOR.
BIRCH	EAST OF MISSISSIPPI, NORTH OF GULF COAST STATES, SOUTHEAST CANADA, NEWFOUNDLAND	HARD, DURABLE, FINE GRAIN, EVEN TEXTURE, HEAVY AND STIFF, AS WELL AS STRONG, WORKS EASILY, TAKES A HIGH POLISH. HEARTWOOD IS LIGHT TO DARK REDDISH BROWN.
BUTTERNUT	SOUTHERN CANADA, MINNESOTA, EASTERN U.S., TO ALABAMA, FLORIDA	MUCH LIKE WALNUT, BUT COLOR IS SOFTER, NOT AS SOFT AS WHITE PINE AND BASSWOOD, EASY TO WORK, FAIRLY STRONG.
CHERRY	EASTERN U.S.	SUPERB WORKING CLOSE GRAINED FURNITURE WOOD. REDDISH COLOR, DARKENS WITH AGE IF NOT STAINED. DURABLE, STRONG, EASY TO MACHINE.
CYPRESS	MARYLAND TO TEXAS	RESEMBLES WHITE CEDAR, WATER-RESISTANT, VERY DURABLE, MAY BE EXPENSIVE AND DIFFICULT TO LOCATE.
DOUGLAS FIR	PACIFIC COAST, BRITISH COLUMBIA	STRONG, LIGHT, CLEAR-GRAINED TENDS TO BRITTLENESS, HEARTWOOD SOMEWHAT RESISTANT TO WEATHERING, AVAILABLE, MODERATELY PRICED.
ELM	EAST OF COLORADO	SLIPPERY, HEAVY, HARD, TOUGH, DIFFICULT TO SPLIT, DURABLE.

Useful Woods (cont.)

WOOD	LOCALE	CHARACTERISTICS
HICKORY	ARKANSAS, TENNESSEE, OHIO, KENTUCKY	VERY HEAVY, HARD, TOUGH, STRONGEST AND TOUGHEST OF OUR NATIVE HARDWOODS. CHECKS, SHRINKS, DIFFICULT TO WORK.
LIGNUM VITAE	CENTRAL AMERICA	DARK GREENISH BROWN WOOD, UNUSUALLY HARD, CLOSE GRAINED, EXCEPTIONALLY HEAVY, HARD TO WORK, CHARACTERIZED BY A SOAPY FEEL.
LIVE OAK	COASTS OF OREGON, CALIFORNIA, SOUTHERN ATLANTIC AND GULF STATES	HEAVY, HARD, STRONG, DURABLE. A BEAR TO WORK, BUT SUPERB FOR SMALL PROJECTS OTHERWISE.
MAHOGANY	HONDURAS, MEXICO, CENTRAL AMERICA, FLORIDA, WEST INDIES, CENTRAL AFRICA	BROWN TO RED COLOR, ONE OF THE TOP CABINET WOODS, HARD, DURABLE, DOES NOT SPLIT BADLY, OPEN GRAINED, BUT CHECKS, SWELLS, SHRINKS, WARPS SLIGHTLY.
MAPLE	ALL STATES EAST OF COLORADO, SOUTHERN CANADA	HEAVY, TOUGH, STRONG, EASY TO WORK, NOT DURABLE. MAY BE COSTLY. ROCK, OR SUGAR, MAPLE IS THE HARDEST.
NORWAY PINE	STATES ALONG GREAT LAKES	LIGHT COLORED, MODERATELY HARD FOR SOFTWOOD, NOT DURABLE, EASY TO WORK.
POPLAR	VIRGINIA, W. VIRGINIA, KENTUCKY, ALONG MISSISSIPPI VALLEY	SOFT, CHEAP HARDWOOD, GOOD FOR WIDE BOARDS — TREE GROWS FAST AND STRAIGHT — ROTS QUICKLY IF NOT PROTECTED, WORKS EASILY. WARPS, BRITTLE, FINE TEXTURE.
RED CEDAR	EAST OF COLORADO, NORTH OF FLORIDA	VERY LIGHT, VERY SOFT, WEAK, BRITTLE WOOD, WORKS EASILY, MAY BE HARD TO FIND IN WIDE BOARDS, VERY DURABLE.

Useful Woods (cont.)

WOOD	LOCALE	CHARACTERISTICS
RED OAK	VIRGINIA, W. VIRGINIA, KENTUCKY, TENN., ARKANSAS, OHIO, MISSOURI, MARYLAND, PARTS OF NEW YORK	COARSE GRAINED, EASILY WARPED AND NOT DURABLE. FORGET FOR DOOR USES.
REDWOOD	CALIFORNIA	IDEAL CONSTRUCTION AND DURABILITY CHARACTERISTICS, TENDS TO HIGHER COST, NOT AS STRONG AS YELLOW PINE, BUT SHRINKS AND SPLITS LITTLE, IS STRAIGHT-GRAINED, EXCEPTIONALLY DURABLE WITH NO FINISH AT ALL, MANY INEXPENSIVE GRADES AVAILABLE, AND POSSIBLY SUITABLE.
SPRUCE	NEW YORK, NEW ENGLAND, W. VIRGINIA, GREAT LAKES STATES, IDAHO, WASHINGTON, OREGON, MUCH OF CENTRAL CANADA	LIGHT, SOFT, FAIRLY DURABLE WOOD THAT IS CLOSE TO IDEAL FOR OUTDOOR PROJECTS.
SUGAR PINE	CALIFORNIA, OREGON	VERY LIGHT, SOFT, RESEMBLES WHITE PINE CLOSELY.
WALNUT	EASTERN HALF OF U.S., SOME IN NEW MEXICO, ARIZONA, CALIFORNIA	FINE FURNITURE WOOD, CONSIDERED BY MANY TO BE THE ULTIMATE. COARSE GRAINED, BUT TAKES SUPER FINISH WHEN PORES ARE FILLED, DURABLE, BRITTLE, MODEST SHRINKAGE, OFTEN KNOTTY.
WHITE CEDAR	EASTERN COAST OF THE U.S., AROUND GREAT LAKES	SOFT, LIGHT, DURABLE WOOD, CLOSE-GRAINED, EXCELLENT FOR OUTDOOR USES.
WHITE OAK	VIRGINIA, W. VIRGINIA, TENNESSEE, ARKANSAS, OHIO, KENTUCKY, MISSOURI, MARYLAND, INDIANA	HEAVY, HARD, STRONG, MODERATELY COARSE GRAIN, TOUGH, DENSE, MOST DURABLE OF ALL NATIVE AMERICAN HARDWOODS, REASONABLY EASY TO WORK (WITH SHARP TOOLS), TENDENCY TO SHRINK, CRACK, MAY BE COSTLY IN SOME LOCALES.

Useful Woods (cont.)

WOOD	LOCALE	CHARACTERISTICS
WHITE PINE	MINNESOTA, WISCONSIN, MAINE, MICHIGAN, IDAHO, MONTANA, OREGON, WASHINGTON, CALIFORNIA, SOME STANDS IN EASTERN STATES OTHER THAN MAINE	FINE GRAINED, EASILY WORKED, SOMETIMES FOUND WITH FEW KNOTS, DURABLE, SOFT, NOT EXCEPTIONALLY STRONG, ECONOMICAL, EXCELLENT FOR MANY USES. WHITE IN COLOR, SHRINKS, DOES NOT SPLIT EASILY.
YELLOW PINE	VIRGINIA TO TEXAS, SOME SPECIES CLASSED AS SOUTHERN PINE	HARD, TOUGH SOFTWOOD, HEARTWOOD IS FAIRLY DURABLE, HARD TO NAIL, SAWS AND GENERALLY WORKS EASILY, INEXPENSIVE, EXCELLENT FOR OUTDOOR USES. GRAIN VARIABLE, REDDISH BROWN IN COLOR, HEAVY FOR A SOFTWOOD, RESINOUS.

In most cases, it does not pay to shop for lumber lower than Select grade, though for small projects, if you can pick and choose, the top two common grades, No. 1 and No. 2, may provide somewhat cheaper possibilities. For larger hardwood projects, FAS is almost imperative.

You will have to join several narrow boards edge to edge to create wide boards for projects that require them, unless you are both lucky and rich, which is why so much of our later attention in this book is paid to various methods of edge-joining boards. Edge-joined wide boards tend to be more stable than single piece wide boards.

Defects in Wood

Examining lumber for purchase, you will note some defects that are obvious. Some may not become obvious until later, as you may find internal pin knots that don't appear on the surfaces.

Wane is bark along the edge of the board, or missing wood along the edge of the board, usually caused by bark dropping off (Figure 1-6).

Checking is splitting of the board, usually at the ends, but sometimes at other spots (Figure 1-7).

Cupping is warping across the width of a board.

Warping is any distortion of the shape of the wood.

Crook is a form of warping, a deviation from end to end straightness.

Crack is a large radial check.

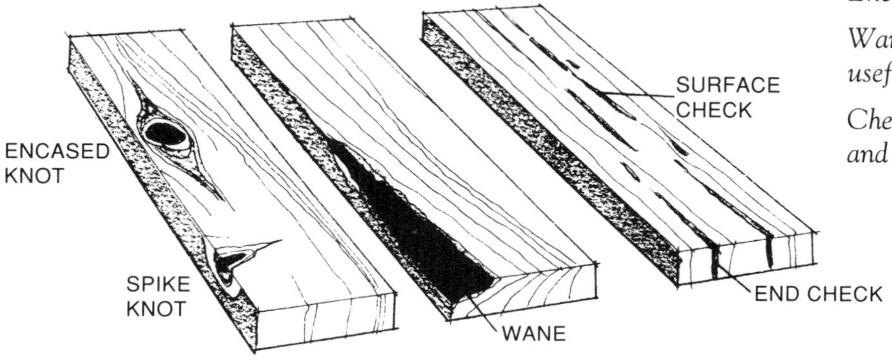

Figs. 1-5 – 1-7.

Encased and spike knots.

Wane on a board edge reduces its useful width.

Checking is found on the surface and ends of boards.

ENCASED KNOT

SPIKE KNOT

WANE

SURFACE CHECK

END CHECK

Fig. 1-5   Fig. 1-6   Fig. 1-7

Diamonding is a form of warp, too, when a section of a board twists clockwise or counterclockwise, and thus appears as a diamond.

Sap stain is a bluish stain caused by fungi in wood, and on the surface of the wood.

Encased, or black, knots are knots that are loose, but remain in the tree, trapped by later growth (Figures 1-5, 1-8).

Fig. 1-8 Encased knots often disappear when a board is handled or worked into a project.

Knots are parts of branches which the expanding tree has overgrown. *Pin knots* are less than a quarter-inch in diameter, while a *spike knot* has been cut along its long axis, giving its exposed section a stretched appearance (Figures 1-9, 1-10, 1-11).

Solid knots are solid throughout the board and show no signs of rot or looseness.

PLYWOOD

Plywood came into its own during World War II, and has taken more of a bad rap than it deserves. Most problems with plywood are caused by using inappropriate types

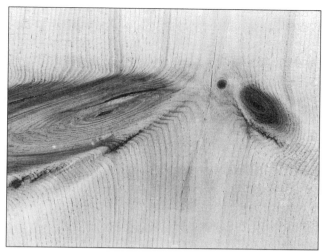

Fig. 1-9 Spike knot, pin knot, and sound knot in one board.

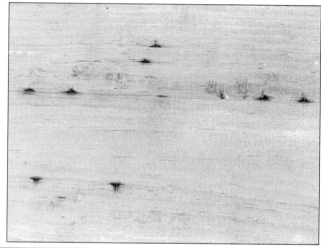

Fig. 1-10 Pin knots.

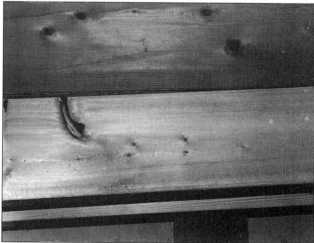

Fig. 1-11 Severe spike knots.

for the jobs being done, taking too little care in joint-making, and improper finishing. In addition, there now is more than a little junk plywood floating around on the market. The big sheets can also be hard to handle, causing problems if the shop is not equipped to handle such sheet sizes.

Softwood plywoods are graded according to standards set by the American Plywood Association for plies, adhesives, and face grades. The actual grading standards would take pages of this book, and are not necessary to our purposes. There are two general adhesive types, Interior and Exterior, with the glues used suitable for the specified use. Interior plywoods have glues that will resist moisture in moderate amounts. Exterior plywood glues resist moisture, period. Marine plywoods, a third and more costly type, resist just about any kind of immersion in water.

Remember that the wood itself moves around quite a bit when moisture levels change, so plies of any type may separate after enough wettings.

Plywood is laid up in odd-numbered plies, with ⅛" and ¼" sheets having three plies, ⅜" and ½" thicknesses having five plies, and so on. Each ply thickness changes as plywood thickness changes. For most uses, you will seldom see more than seven plies, though some types have as many as fifteen. Sheets, or panels, are available in 4' widths with lengths ranging from 8' to 12'. Some may be special ordered longer.

Various wood species are used for softwood plywood interior plies. Some is softwood, some hardwood, though usually it's aspen, the softer maples, cottonwood, basswood, and similar hardwoods, and some may be beech, birch, sugar (hard or rock) maple, lauan, and sweet gum. The lower the group number — there are five — the stiffer and stronger the plywood, so look for Group 1 or 2 when strength and stiffness are important.

Face grades of softwood plywood range from A down to D, with A being the best; grade N veneers are special order furniture grade types. Both faces are graded, as in

A-A, A-C, etc. Plywood also comes in engineered grades and appearance grades, with the engineered grades being the strongest and the appearance grades the prettiest.

Even cheap hardwood plywood tends to cost double what typical A-C or A-B sanded grades cost, but offers rotary or sliced veneers over a particle board, veneer, or lumber core center. Hardwood face grades are identified; if you order cherry hardwood plywood, cherry is the face veneer on both sides.

Premium Grade A or #1 must match multiple veneer pieces if they are used to make the face, so their figure display appears as closely matched as possible. Contrast in color and grain is avoided. Special combinations can also be ordered (Specialty Grade, or SP). Good Grade (#1) hardwood plywood has its face veneer so placed as to avoid sharp contrasts of color and grain, while Sound Grade (#2) has unmatched face veneer grain and color. Hardwood plywood comes in three adhesive grades: Type 1 and Technical Type are manufactured using waterproof adhesives, made to withstand repeated moisture exposure without delamination, while Type II has a moisture-resistant adhesive and is used where repeated wetting is possible, but where constant wetting is not a problem. Type III adhesives are for materials intended for rough uses such as in packing cases, crates and similar items.

Many other wood composite boards are available. Check your needs against costs, and wander through lumberyards and mills to see what they have or are making. Sending off for catalog material and brochures from major lumber manufacturers will also help you keep abreast of the trends, if retail suppliers in your locale do not keep such material on hand. The basics of wood do not change, but methods of using once less-than-useful bits and pieces, methods of producing manufactured woods, and types of manufactured woods do.

Know the Pitfalls

That's the end of the horror story. Selecting wood is not really as bad as all that sounds, nor is caring for it after it is selected, but it is best to know the pitfalls, so you can avoid them. There are certain tenets to follow when selecting wood, particularly at today's high prices. Knowing what you're buying is the best of them. Plan carefully as to project needs so waste is as low as possible, and then do your best in selecting the wood at the supplier.

If your supplier doesn't let you paw through the wood piles, you will need some time to see how much of your interest he or she has at heart. If you are constantly getting wood that contains too much waste, complain. If that does no good, find another supplier, preferably one who will let you make your own selection. Woodworking is far more enjoyable when you work with good tools and good wood, with the major portions of waste coming from your own mistakes.

CHAPTER 2

WOOD JOINTS AND WHAT THEY DO

S electing the wood joints for a project depends on the strength and appearance needs of that project. Several joints may serve better than using a single joint style all the way through a project.

A good understanding of the fundamentals of basic joint types will give you a stepping-off point that allows you to select joints to suit project design. That grasp of fundamentals allows you to determine and make joint design changes.

In general, woodworking joints fit into one of two categories. A joint is either for carcass (larger, flat wood members) construction, or for framing (smaller, thinner flat wood) construction (Figures 2-1, 2-2).

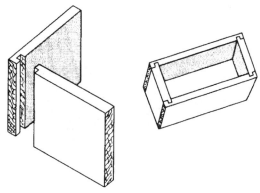

Fig. 2-1 Dado and rabbet are used in carcass construction.

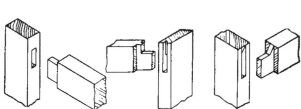

Fig. 2-2 Stub, haunched, and table-haunched mortise-and-tenon framing joints.

CARCASS JOINTS

Butt joints are the simplest of all carcass joints. Butt joints bring two flat pieces of wood together to form a junction, which may be an L, a T, or a flat board. Common uses include all sorts of framing (with large timbers, the butt joint becomes a framing joint), producing wide, flat boards, and joining carcass pieces in simple cabinets, chests, and other projects.

Butt joints are often splined to increase strength and ease alignment problems. Splining requires a groove in both pieces, the insertion of the spline, and gluing. (Most butt joints are glued for cabinet construction; in heavier work, they are nailed or screwed.) Developments in splining include biscuit joinery: using short splines shaped like flat footballs to make strong butt joints in any of the styles. Biscuit joiners make alignment of the splines easy, and installation of the biscuits is quick. The biscuits swell when glued and aid strength and alignment (Figure 2-3).

Fig. 2-3 Porter-Cable's biscuit joiner, and biscuits.

Miter Joints

Miter joints are angled butt joints. While most people think of door frame and picture frame molding when miter joints are mentioned, they are useful in other ways. They provide the simplest method of joining plywood so that plies don't show. Miter joints may be splined or biscuit joined.

Square miter joints form a 90° angle, with each member cut at 45°. There are many variants, some with 60° angles joined to 30°, and others with different angles of cuts. To form a true miter, the angle cut is one-half the angle joined.

Stopped miters are sometimes used when one piece is thicker than the other, with both pieces visible. The larger piece has the miter stopped and a straight cut made from that point, while the smaller piece is mitered and fits into the miter and lip thus created.

Miter joints are often keyed. That is, a slot or other shape is cut through both pieces of the joined miter, and a piece to fit the cutout shape is inserted to aid strength — and often appearance. Most such keys are simple splines, added from the outside, and are called slip keys. Dovetail keys are possible, if you wish to do the work; they are great when you want to contrast woods as well as add strength.

Tongue-and-Groove Joints

We're all familiar with tongue-and-groove joints, in appearance if nothing else. That is how most of our wood flooring is joined. The tongue and groove is a simple joint, made easily on table saws with molding heads, with routers, and on shapers. The groove is cut in one piece, while the tongue is cut on the other side. Pieces are then assembled one to the other, tongue to groove, and nailed, through the tongue and into the joist, to hold the floor in place. Tongue-and-groove joints are excellent alignment aids for producing large, glued flat surfaces from narrow

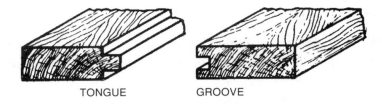

Fig. 2-4 A tongue-and-groove joint. Courtesy of Stanley Tools.

TONGUE   GROOVE

boards. They also aid strength by adding glue surface (Figure 2-4).

Dovetails

Probably the least popular joint for novice wood-workers, the dovetail is difficult to make properly by hand. Dovetail joints form shapes like a bird's tail, narrow at one end and fanning out in width as the wood reaches its endpoint. Until development of router jigs for dovetail-ing, it was a dying art form because of the difficulty of making them. It had long been replaced by machine-cut joints in most furniture because of costs. The dovetail, though, has made a very strong comeback in recent years.

The through dovetail joint shows the end grain of both pins and tails. The tail is the portion that gives the joint its name, looking a lot like a spread bird's tail. The pins are the smaller pieces that are present on each side of the tail. Half blind, or lap, dovetails show the end grain of the pin pieces only. Mitered dovetails (called secret dovetails in a more dramatic age) show no end grain. Within these three basic categories are a number of sub-categories, some useful, many not (Figure 2-5).

The through dovetail is a showy joint, and, if well cut, adds to project appeal. It is useful for showing off combi-nations of light and dark woods. Any dovetail resists both pull and draw forces, so is great in drawer construction, with through dovetails most commonly used on drawer backs. The through dovetail is also useful in general carcass framing for chests, where you wish joint design to be obvious. Hand-cut or large template dovetails are used for larger chests.

For drawer fronts, the half-blind dovetail resists both

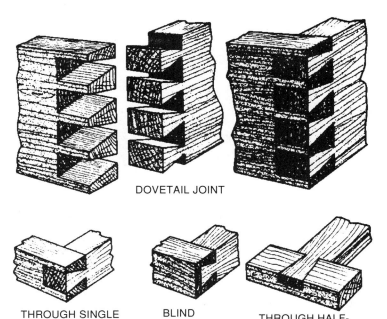

DOVETAIL JOINT

THROUGH SINGLE
DOVETAIL

BLIND
SINGLE
DOVETAIL

THROUGH HALF-
LAP SINGLE
DOVETAIL

Fig. 2-5 Through dovetail (top), through single dovetail (bottom left), blind single dovetail (center), and through half-lap joints.

pull and draw forces, but does not show the joint style to the world. It is a fine joint for more formal pieces, where it was traditional until recent years. The half-blind dovetail is an oddity, a joint very difficult to cut by hand, but one you may cut readily with almost any dovetailing jig for routers.

Mitered dovetails are seldom used today, and were seldom used in the past. Utility is the same as for other dovetails, but layout and cutting is absurdly complex. To my knowledge, there is no power tool jig that reproduces this joint with any degree of ease.

Larger single dovetail joints are sometimes used for framing. These are most often through dovetails, also known as half-lap joints. A single member is cut as a tail, with a cutout in another frame rail showing two pins to the outside of the tail (Figure 2-6).

The strength of the dovetail is most appreciated by those constructing drawers, chests, and cabinets where

Fig. 2-5, cont.

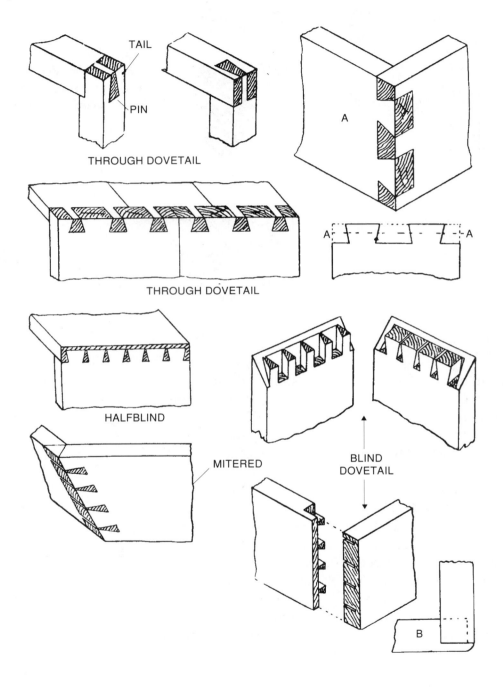

TAIL

PIN

THROUGH DOVETAIL

A

THROUGH DOVETAIL

HALFBLIND

MITERED

BLIND DOVETAIL

B

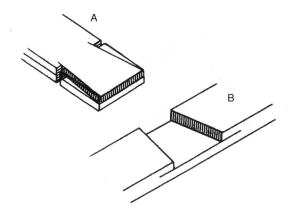

Fig. 2-6 *Making a dovetail half-lap joint.*

stress may come from several directions. The dovetail joint, properly cut, requires little glue to resist this multi-directional stress.

Dovetails may be cut by hand, if you have sufficient time and the desire to lay them out (Figures 2-7, 2-8) or by machine, using any of a number of dovetailing jigs and a router (see chapter 3). Dovetails may also be cut on the bandsaw. Rough dovetails for framing and heavy timbers are sometimes cut with jigsaws.

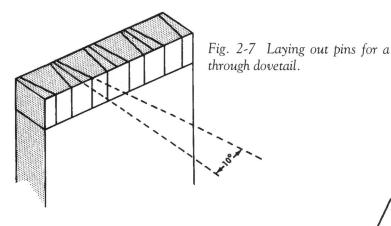

Fig. 2-7 *Laying out pins for a through dovetail.*

Fig. 2-8 *Chiseling out the waste in a through dovetail.*

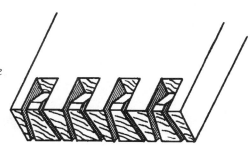

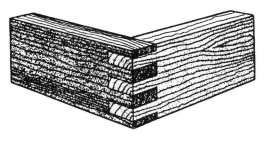

Fig. 2-9 *Finger corner joints, also called box joints.*

Box (Finger) Joints

The finger, or box, joint appears much like the dovetail joint. It is used in the same way, for the same types of projects, but is simpler to lay out and cut. The fingers are usually cut with a table saw or router (Figure 2-9).

The finger joint is stronger than the dovetail joint when used in boxes, drawers, chests, and other carcass construction because of the huge glue area it supplies. The box joint has replaced the dovetail in much fine furniture in the past century. Joint variations abound.

The knuckle finger joint is an example, offering a hinge knuckle (the cylindrical part of a hinge where the two pieces join), cut from a finger joint, with a dowel inserted as a pivot pin. The fingers are rounded to allow hinging. Finger joints are more easily cut on mitered surfaces than are dovetails, and the joint is easier to match up and produce with rapidity, without the need for a great deal of practice, beyond learning the basic use of a table saw.

The usual method of producing finger joints is to make a jig (see chapter 4) that gives the needed width cuts with a dado blade, with a spacing bar setting the distance between finger slots. The jig is simple to make and use.

One commercial jig exists. The Accu-Joint works to produce three sizes of finger joints, on boards up to 12" wide. The jig is simple to use, and requires only a backing board for the table saw miter gauge, for each size of joint.

The Porter-Cable Omni-Jig (see chapter 3) offers a template that will let you produce finger joints using a router. This is simple to use, though costly and limited to ½" finger joints. To ease the impact of the cost, you can remember that the Omni-Jig also aids in the production of numerous styles of dovetail joints.

Corner Locking Joints

This is a machine-made joint that is simple to produce with a router and special bit. It isn't as pretty as the dovetail, but is much faster to produce. It may also be cut on the table saw, where it is more complicated to make, and requires great cut accuracy. It can be quite enjoyable to produce, and is fun to practice.

FRAMING JOINTS

For most cabinetry, the mortise and tenon is the primary framing joint. Tenons are the parts that fit into the mortises, or holes, cut in other frame members. For joining cabinet and other frames, and for joining such things as furniture legs to stretchers and aprons, there is nothing stronger than a good, tightly cut mortise-and-tenon joint.

There are almost innumerable variations on the mortise-and-tenon themes, largely because the joint is so useful and strong in holding large pieces together (Figure 2-10).

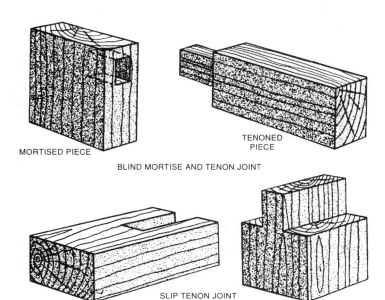

MORTISED PIECE

TENONED PIECE

BLIND MORTISE AND TENON JOINT

SLIP TENON JOINT

Fig. 2-10 Mortise-and-tenon (blind) and slip-tenon (bridle) joints.

Most tenons have shoulders (the wood from which the tenons extend) on four sides of the piece, though not all do. Shoulders may be on one side, two sides, or three sides as well, depending on construction needs. Common use indicates a tenon about one-third the thickness of the members being joined, if both members are the same thickness. This leaves enough material on the mortise side to help keep twisting forces from snapping the tenon through the mortise side wall. No single tenon should exceed, in width, six times its thickness. That means, for example, that a ¼" wide tenon shouldn't be more than 1½" long for maximum strength.

There are a number of different forms of mortises and tenons. A stopped, or blind, mortise, for example, is one that does not go through the mortised wood. The stopped mortise is matched with a stub tenon. The through mortise takes a through tenon, exposing the end grain of the tenon. This may also be foxed, or wedged, into its end. A stub tenon may also be wedged. A tenon that is wedged spreads against the mortise sides. The best protection against pull-out is a slightly spread mortise to accept the wedged end of the tenon, whether it is a stub or through tenon. The gunstock tenon adds a slight curve to a shoulder of the tenon so that it matches the design of a curved rail and provides a continued design arch. Often it is necessary to design a supporting notch on the mortise member so that there is no unsupported piece, in the arch, with short grain. Such short grain pieces are weak and break off easily.

Doubled mortise and tenons in a single joint do not produce greater strength, but are sometimes used when material requirements demand such uses. Experienced joiners look on doubled mortise and tenons as necessary evils, because two thin tenons are not equal in strength to one thick one.

Haunched mortise and tenons are specialized joints sometimes used on the outside members of a frame. They are a tenon with a section cut away to allow resistance to wedging forces, while the mortise is left filled in with the area that the tenon was cut away from. Usually, a short stub, or haunch, is left on the tenon, which is received in a special stub, or haunched, mortise. With the haunch down, the resistance to twisting forces is very high for tables, and similar structures (Figure 2-11).

Other Framing Joints

Thin frames, such as those of cabinets, may be joined with other than mortise-and-tenon joints, though tradition says stick with the mortise-and-tenon types. In fact, some of the joints used are mortise-and-tenon joints with slightly different names.

For example, a bridle joint is a mortise-and-tenon joint that has three sides of the mortise open, thus forming a "bridle." That usually means the tenon only needs two sides trimmed to fit, and the joint is a simple slip fit. The term bridle joint is, of course, related to a horse's bridle and its open end, thus the outer, or mortise, section of the joint bridles the tenon, holding it in place.

The common description of a bridle joint is that it is a reversed mortise-and-tenon joint (Figure 2-12). It isn't. Nor is it a mortise-and-tenon joint that has had the wood normally left solid in a mortise cut away. It is a variant of the mortise and tenon, usually with one piece of the mortise cut away to allow the bridle to form. The bridle joint is not as strong as a mortise and tenon, but is stronger than lap and half-lap joints, and is useful in light framing.

Other joints are simply versions of mortise and tenon with changes. A splined and doweled butt miter is a good example. Here the miter is made, and the joint checked for fit. Then the center thirds of the miters are removed, as if each were an open mortise joint.

The length of material, or depth, removed at the back of the mortise is determined in a simple manner. Measure down the cut face of the miter, and divide by five. Measure below the miter heel that one-fifth, and mark. Carry straight across the board with a square. That provides the depth of cut for removing the mortise. Once the

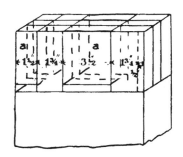

Fig. 2-11 Proportions for tenons.

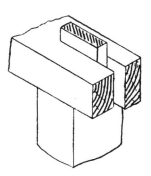

Fig. 2-12 Bridle joint.

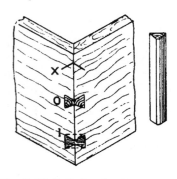

Fig. 2-13 Splining for appearance in a miter joint.

spline is cut and inserted, you can drill the dowel holes. This may be done in two steps, before the spline is inserted, or after the spline is inserted.

Fitting such splined miter joints may prove difficult. Check the joint as the spline is being fitted, and plane the spline in small steps (Figure 2-13).

Splined joints can be designed and repeated around any basic corner joint, and will result in quite a fancy look if different colored woods are used as keys. The keys may be cut in different shapes as well, giving an even fancier look.

LENGTH JOINING

It is no longer essential to add great mechanical strength to length joinery, because of modern glues. Still, the joint designs remain, and may be of interest to some people at some time.

Scarf joints and the variations on them are the primary methods of joining for length without using glue. Of course, a simple scarf joint is nothing more than an angled slice taken out of a board's end, with the grain, matched with a similar cut on another board end. This provides good glue surface area for modern use, but didn't work for our ancestors who lacked high-strength glues.

Splayed scarf joints, either doweled or keyed (and the key doweled from the sides), were extremely useful at one time and had a number of variants. The joint is used to join timbers of the same size. The angle of the scarf works off a length-times-width ratio of 2½:1. That is, the length of the joint is 2½ times its width.

The ends of the joint are set in ³⁄₁₆" from the top and bottom surfaces. Those two points are then connected, which gives the angle of splay. The line is drawn, and the center butt end lap is lifted off at a 60° angle. Once the layout is done, this hand-cut joint is cut. Bring saw cuts down to the lines before chiseling out waste.

The key for the mortise is marked out last, with the pieces temporarily fitted together. The mortise is keyed, or tapered, and the side lines on the pieces are kicked apart just slightly so the key provides compressive pressure when it is driven into place. The key may then be pegged, or doweled, from the sides.

Different versions of the splayed scarf exist, including those pegged from the top, and with all kinds of keys. Some are simpler, most are more complex, and their uses today are extremely limited.

Dowel Joiner

One of the most difficult and overrated types of joinery is doweling. Dowels can provide good support and help for a number of joints, but add complexity to what are usually simple joints, and may present alignment problems that are more difficult to handle than the more difficult joints would be. Dowels are substitutes for more complex, harder to cut joints such as mortises and tenons, and are not as good.

Too, dowels tend to present gluing problems. The need for gluing grooves and that hair of extra depth needed in the drilled holes are things some woodworkers forget, much to their chagrin when the joint fails early.

Dowels, to be examined in more detail later in this book, are virtually all made of hardwood rod, often birch, turned to exact diameters. Many are now grooved to allow glue flow, and some even come with beveled ends. If you want off sizes or lengths, you'll have to form them yourself (Figure 2-14).

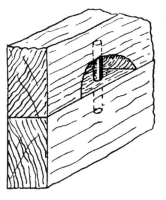

Fig. 2-14 Doweled joint.

Biscuit (Plate) Joinery

The biscuit joiner is doing a lot to replace dowels these days. These flat, football-shaped pieces of compressed wood give a means of aligning and supporting joints that is, for most of us, far simpler than fooling around with dowels and doweling jigs. The small saw blade in the plate, or biscuit, joiner slices slots on both sides of a joint, the biscuit is fitted in after glue is applied,

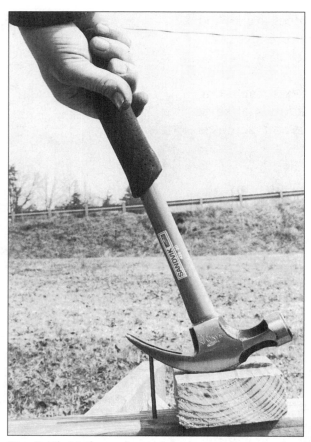

and the joint comes together, with alignment needed only in one plane.

For most uses where the available stock is large enough — the slot requires about 2¾" for even the smallest biscuit — biscuit joinery is replacing both butt and dowel joinery.

WAYS TO FASTEN JOINTS

This chapter has provided a look at some of the many woodworking joints that may be used in projects. As we move through the book, we will see how each of these types is cut — and how many different ways there are to cut some joints.

There are dozens of variants of most woodworking joints, and nearly as many ways of fastening the joints to prevent separation. Glue is the standard for most kinds of furniture work, but is not a solution for items that need to be taken apart. Nor does glue work well where joints must fit more loosely than is customary for furniture. Such loose-fitting joints are used in outdoor furniture, where wood movements are more extreme in shorter periods of time. Speed of joinery also determines the type of fasteners. Nails are used for house framing instead of fancy joints, screws, or glues, because they are faster and cheaper (Figure 2-15). Screws allow pieces to be taken down and moved without destruction, while nails do not. Specific fasteners, called knock-down fasteners, make take down and movement of furniture and other

Fig. 2-15 Nails can be used to put a project together quickly, but changes and mistakes are not always easily corrected without damaging materials. This technique is best for removing nails: use a block to get the hammer off the work surface.

items easier, while providing a more solid joint than simple screws.

Joinery combines joint design with fastener or adhesive type to best suit a particular purpose. Care in design, cut, and assembly assures long life for any project. One of the prime cares needed is in measuring and marking.

MEASURING AND MARKING

Always aim for top quality. This is the result when joints are accurately marked and cut, to length, width, and thickness.

Those requirements lead naturally to the use of measuring and marking tools. It matters not at all whether you use power or hand tools: if a joint is not measured and laid out accurately, it is not going to work well or long. Thus you must work with good quality folding rules, measuring tapes, and measuring jigs.

Too, the use of marking tools should not be confined to the carpenter's pencil. Utility knives and scribes are much better for precise work. Incising marking tools provide a clean, narrow line for far greater accuracy (Figure 2-16).

There is a variety of marking tools for specific jobs, most of which scribe lines. These include marking gauges and mortise gauges. Such tools are handy, but not essential. Poorly made ones are worse than nothing (Figures 2-17, 2-18, 2-19).

Try squares must be checked for accuracy. For precise work, combination squares are not recommended as they fall out of square easily. To check any square, mark a board with it (this assumes you know the board has parallel sides), then reverse the square and mark the board from the other side to the same original mark. If the lines match, or are parallel if drawn at a slight distance from each other, the square is fine. If the lines are not right, adjust or replace the square.

Squares are precision instruments, but are adjusted roughly. Set the square on a solid surface, handle down.

Fig. 2-16 Using a square. Courtesy of Stanley Tools.

1. WORK FACE

PLANE ONE BROAD SURFACE
SMOOTH AND STRAIGHT. TEST
IT CROSSWISE, LENGTHWISE,
AND FROM CORNER TO CORNER.
MARK THE WORK FACE X.

2. WORK EDGE

PLANE ONE EDGE SMOOTH, STRAIGHT
AND SQUARE TO THE WORK FACE. TEST
IT FROM THE WORK FACE. MARK THE
WORK EDGE X.

3. WORK END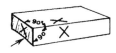

PLANE ONE END SMOOTH AND SQUARE. TEST IT FROM
THE WORK FACE AND WORK EDGE. MARK THE WORK
END X.

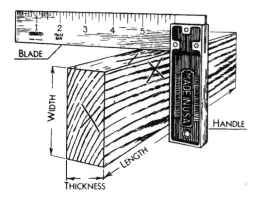

HOLD THE HANDLE OF THE TRY SQUARE
TIGHT AGAINST THE STOCK WHEN TESTING
ENDS, EDGES, OR SCRIBING LINES.

Fig. 2-16, cont.

4. SECOND END

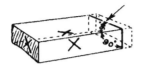  

MEASURE LENGTH AND SCRIBE AROUND THE STOCK, A LINE SQUARE TO THE WORK EDGE AND WORK FACE. SAW OFF EXCESS STOCK NEAR THE LINE AND PLANE SMOOTH TO THE SCRIBED LINE. TEST THE SECOND END FROM BOTH THE WORK FACE AND THE WORK EDGE.

5. SECOND EDGE

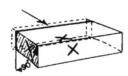 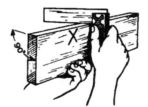

FROM THE WORK EDGE GAUGE A LINE FOR WIDTH ON BOTH FACES. PLANE SMOOTH, STRAIGHT, SQUARE AND TO THE GAUGE LINE. TEST THE SECOND EDGE FROM THE WORK FACE.

6. SECOND FACE

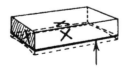 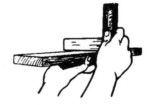

FROM THE WORK FACE GAUGE A LINE FOR THICKNESS AROUND THE STOCK. PLANE THE STOCK TO THE GAUGE LINE. TEST THE SECOND FACE AS THE WORK FACE IS TESTED.

Fig. 2-17 Using a marking gauge. Courtesy of Stanley Tools.

THE THUMB SCREW

NO ROLL MOTION IS NECESSARY

THUMB SCREW

SHOE

PIN

FACE PLATE

BEAM

STOP SCREW

HEAD

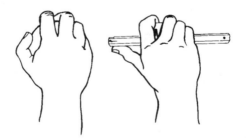

SET THE MARKING GAUGE BY MEASUREMENT
FROM THE HEAD TO THE PIN. CHECK THE
MEASUREMENT AFTER TIGHTENING THE THUMB
SCREW.

HOLD THE GAUGE AS YOU WOULD A BALL.
ADVANCE THE THUMB TOWARD THE PIN SO AS
TO DISTRIBUTE THE PRESSURE EVENLY
BETWEEN THE PIN AND THE HEAD.

THE PIN SHOULD PROJECT ABOUT $1/16$ IN. THE
CURVED SIDE OF THE PIN HELPS TO KEEP IT
FROM FOLLOWING THE GRAIN OF THE WOOD.

Fig. 2-17, cont.

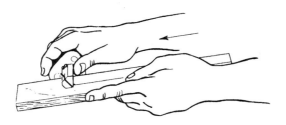

LAY THE BEAM FLAT ON THE WOOD SO THE PIN DRAGS NATURALLY AS THE MARKING GAUGE IS PUSHED AWAY. NO ROLL MOTION IS NECESSARY. THE PIN AND LINE ARE VISIBLE AT ALL TIMES.

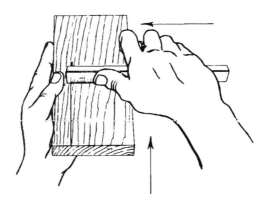

TO MAKE A GAUGE LINE PUSH THE GAUGE FORWARD WITH THE HEAD HELD TIGHT AGAINST THE WORK EDGE OF THE WOOD. THE PRESSURE SHOULD BE APPLIED IN THE DIRECTION OF THE ARROWS.

THE PIN IS GROUND WITH A CONICAL POINT THEN ONE HALF IS GROUND FLAT. THIS GIVES A KNIFE TYPE LINE.

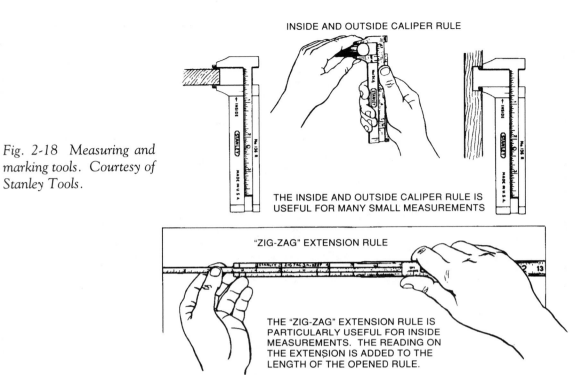

Fig. 2-18 Measuring and marking tools. Courtesy of Stanley Tools.

INSIDE AND OUTSIDE CALIPER RULE

THE INSIDE AND OUTSIDE CALIPER RULE IS USEFUL FOR MANY SMALL MEASUREMENTS

"ZIG-ZAG" EXTENSION RULE

THE "ZIG-ZAG" EXTENSION RULE IS PARTICULARLY USEFUL FOR INSIDE MEASUREMENTS. THE READING ON THE EXTENSION IS ADDED TO THE LENGTH OF THE OPENED RULE.

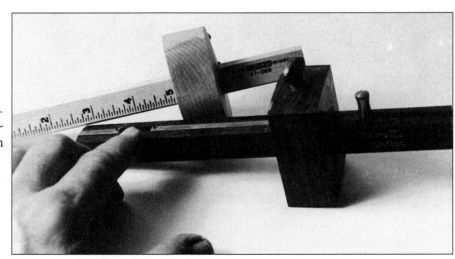

Fig. 2-19 Marking devices. The Rabone in the foreground is double pinned on side for marking mortises.

Tap the blade with a soft-faced hammer in the direction that will correct the fault. Check. Continue tapping until the square is accurate. If the square does not hold its setting well, replace it as soon as you can (Figures 2-20, 2-21).

Measuring tools also need to be checked for accuracy, where possible. We can never be sure which of the checking tools is the accurate one, though. The best

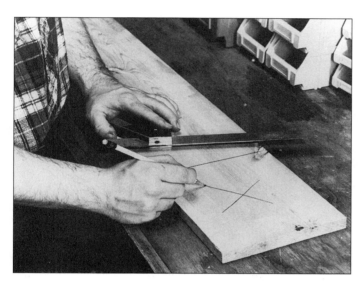

Fig. 2-20 Mark carefully, and then mark waste to prevent errors.

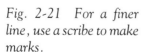

Fig. 2-21 For a finer line, use a scribe to make marks.

method to get around this problem is to use two or three tapes, rules, and folding rules that have been matched to each other. Any deviation, then, will always be the same. With quality measuring tools, the deviation will be small.

CHAPTER 3

Making Joints with a Router

ne of the most amazing tools we have for making joints is the router, once a rare tool, but now commonly found in the workshop.

This tool is precision-built and is similar to a drill in that it has a high speed motor spinning a chuck that will hold any one of many cutting bits. All of this is held in a frame that is part of the base. This base rests on the wood surface, the bit projecting beyond it. The depth of the cut made by the bit is controlled by raising and lowering the motor in the frame. This is done by loosening a locking ring, then turning an adjusting ring that links the motor and the frame.

It is possible to learn the basics of using this tool in minutes. Wood being worked on should be held firmly in place, so the router can be operated with both hands. Clamps are useful for this.

If you do not have a router, but plan to continue woodworking, buy one. Select a model with at least 1 horsepower, preferably with a ½" collet chuck (the part that grips the bit shank). The ¼" collet chuck is less accurate.

A plunge router allows the user to plunge the bit into the workpiece without moving the router base. This means work can be started without tipping the bit into the work, or running it in from the side. The plunge router is handy, but far from essential. Many of the best routers do not offer a plunge feature. Under heavy use, plunge routers wear out a tad sooner. This is not a problem for non-production work, unless the router is a cheap one.

Fig. 3-1 The lightest router I own, this Craftsman shown here, is easy to use, and has its own work light, plus a shaft lock. It is not meant for heavy duty work, but its low cost offsets that.

Fig. 3-2 Porter-Cable's basic, and more or less classic, 690 model here has a D handle that aids control in some uses, for some people. I like the D handle when freehand routing, but find it hangs up too often on most homemade jigs. The 1½-horsepower design has been around for a long time, is durable, plenty powerful for most work, easy to handle, and reasonable in cost. Recently, Porter-Cable has introduced a plunge base for this model which adds to its value.

Routers on the market today offer more features and power than ever before. Power ranges from ½ to 5 horsepower. For most home workshop uses, a quality router offering 1 or more horsepower will serve for years (Figure 3-1). Lighter duty routers are too limited for some work, while the heavier duty models, those of 3 horsepower and more, are heavier and harder to handle, though they can turn out heavy work (Figures 3-2, 3-3, 3-4, 3-5).

Safety is important: there is no guard on the bit, which turns from 15,000 to 30,000 revolutions per minute. Change bits or work on the bit only when the router is unplugged. Let the bit reach full speed before starting work, and let it stop on or in the work before lifting the router.

Routers can produce many types of joints, especially when used in combination with jigs (Figure 3-6). Among

Fig. 3-3 Makita offers this top of the line model with a plunge mechanism that's easy to operate, and a ½" collet. The Porter-Cable offers a choice of ¼", ⅜", or ½" collets. The Makita requires the use of a sleeve to fit smaller shanked bits. It's fairly high in cost, but truly durable under adverse conditions. Plunge routers tend to wear out their plunge mechanisms first, if the bearings around the collet don't go.

Fig. 3-4 The large Ryobi. Both it and the Makita are heavy routers, harder to handle than less powerful models, and not for the beginner. The Ryobi offers a lot of power and a variable speed feature not found in many routers. Durability is excellent, but cost is also tops, as with the Makita. Both routers, and for that matter, most power tools today, are sold at heavily discounted prices.

Fig. 3-5 Dremel's Moto Tool added to its shaper table combine to produce a fine working little tool for those with small work to be done. Dremel has concentrated in this field seemingly forever, and does exceptionally well. A router setup for the Moto-Tool can also be found.

the most popular are dovetail joints and mortise-and-tenon joints. One jig enables you to do finger joints with the router. Lap joints are easily made, as are dadoes and rabbets (Figure 3-7).

Joints

The dovetail is the strongest woodworking joint for most cabinetry use. It is great for drawers and box construction, up to and including large chests, if you have a jig that can produce wide enough sections for carcass joinery (one jig works on any width board). One jig also works to produce sliding dovetails for locked-in shelves, giving strength and stability in bookcases and cabinets.

A number of inexpensive router template jigs on the market allow reasonably rapid cutting of dovetails. These generally are difficult to set up — all commercial jigs are,

Fig. 3-6 *This router is cutting dovetails, using the Omnijig from Porter-Cable.*

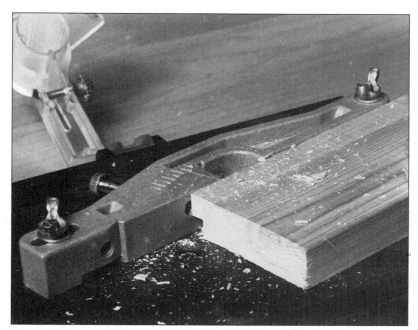

Fig. 3-7 *Cutting a rabbet with the Dremel shaper table setup.*

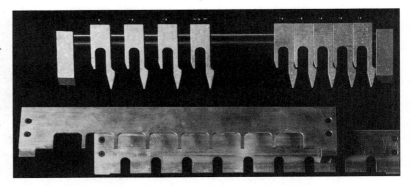

Fig. 3-8 *Here are two types of templates for the OmniJig. The top one is the variable spacing pin type, while the others work with wide spaced dovetails to produce a hand-cut look.*

whatever the claims and type, but the cheaper ones are worse. They are prone to slippage and quick wear. Their dovetailing styles are limited, often to the half-blind dovetail.

The OmniJig

Costly jigs such as the OmniJig offer templates of heavy aluminum that wear for a very long time, making many types and sizes of dovetails possible. The OmniJig will produce half-blind, through, variably spaced, and sliding dovetails, and even finger joints, a feature no other router jig offers (Figures 3-8, 3-9).

Fig. 3-9 *Two of the standard templates for the OmniJig make ½" and ¼" dovetails. There is little difference between these templates and those for cheaper jigs. These are made of machined plate and cast aluminum, while the cheaper ones are of plastic or heavy fiberboard.*

The jig is fussy to set up, but simple to use, once you get the hang of it. Plan on using plenty of scrap stock to get the feel. The OmniJig offers so many choices, time disappears while you're fooling with the machinery. The base is heavy cast iron and machined steel. Templates, which may be added as needed, are either ¼" aluminum plate or ½" aluminum.

The OmniJig is expensive, but it's a tool you'll be leaving to your grandchildren.

The Keller Jig

The Keller dovetail jig takes another approach, offering three sizes of templates, along with carbide bits. While the cost is also high, this rig is the only one to provide unlimited dovetail width. The unit is sturdy, of ½" machined aluminum plate. Templates provide for fixed pin spaced dovetails and come in three versions. For standard uses, the model 2401, at 24" wide, is best. It works with stock ⅜" to 1" thick.

To use this, clamp the stock in a vise, after mounting the template to flat wood blocking. Make adjustments and trial cuts until the fit is right, then cut the tail board. Use the tail board to mark the pin board, then clamp it with its appropriate template and cut it (Figure 3-10).

Fig. 3-10 Keller dovetail templates differ from those used with the OmniJig in that there is no arm at each end. These do not attach to a jig, but form their own with a backing board of wood. One type uses a straight bit, while the other uses a dovetail bit, so that a change of bits is required to cut a set of dovetails.

These two jigs have in common high quality construction, great precision, and durability. They also have aluminum templates. With aluminum templates — or plastic templates, or steel templates — it is essential that the router bit come to a stop to prevent nicking the templates as the tool is raised from the work. If you strike a steel template, the bit may be damaged or metal pieces may fly.

Half-Blind Dovetails

To make half-blind ½" dovetails, useful for attaching drawer fronts to sides, mount the OmniJig on a bench, then insert the template guide in the router base (Figure 3-11). Make sure the router is unplugged.

Fig. 3-11 Accessories needed to cut half-blind ½" dovetails with the OmniJig. This series courtesy of Porter-Cable.

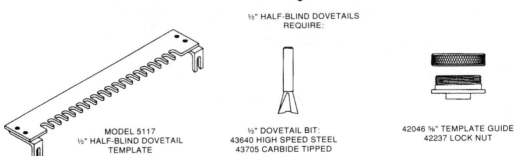

½" HALF-BLIND DOVETAILS
REQUIRE:

MODEL 5117
½" HALF-BLIND DOVETAIL
TEMPLATE

½" DOVETAIL BIT:
43640 HIGH SPEED STEEL
43705 CARBIDE TIPPED

42046 ⅝" TEMPLATE GUIDE
42237 LOCK NUT

Here's a tip for adjusting router bit depth: the router bit, going into the collet, is normally bottomed and then drawn back up ¹⁄₁₆" to make sure the collet catches on the shaft, not on the fillet (the rounded area) under the cutting head. Instead of fiddling with this each time, buy ¼" and ½" O rings. Roll one of these onto the shaft of the bit, up to the spot where the fillet starts. This automatically sets the correct depth.

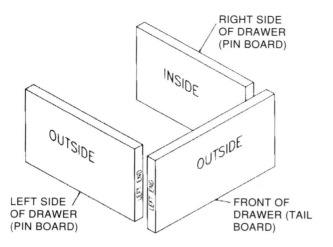

Fig. 3-12 Board arrangement.

RIGHT SIDE
OF DRAWER
(PIN BOARD)

INSIDE

OUTSIDE

OUTSIDE

LEFT SIDE
OF DRAWER
(PIN BOARD)

FRONT OF
DRAWER (TAIL
BOARD)

Adjust the router bit depth according to the manual furnished with your router. Insert the left side of the drawer (the board that gets the pins) inside out (Figure 3-12) under the front clamp, extending it ¼" above the top surface of the jig (Figure 3-13). Place the tail board (drawer front) inside up under the top clamp and butt against the left side of the drawer (Figure 3-14). Place the template on top of the jig, so that there is the correct distance (this varies with different templates and joint designs) from the back edge of the template slots to the edge of the drawer front, at both ends of the template. Tighten the template knobs. Position the drawer front (top or horizontal piece) so that its left edge is past the left edge of the template slot. Clamp the board.

Slide the top stop bar over to touch the drawer front and lock in place with the hex key.

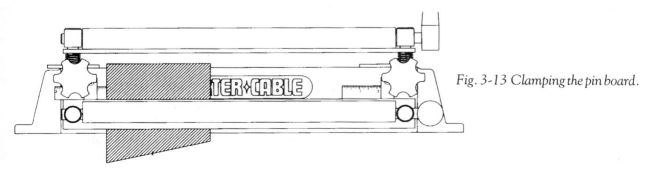

Fig. 3-13 Clamping the pin board.

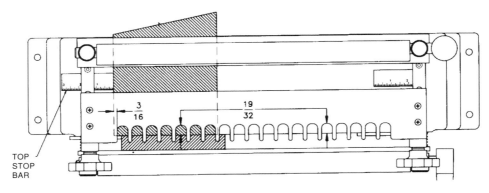

TOP
STOP
BAR

Fig. 3-14 *Aligning the drawer front.*

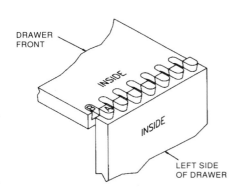

DRAWER
FRONT

INSIDE

INSIDE

LEFT SIDE
OF DRAWER

Fig. 3-15 *Router movement.*

Loosen the front clamp and raise the left side of the drawer (vertical piece) so that it touches the bottom of the template at the specified distance from the left edge of the top board. Clamp. Slide the front left stop bar to the wood and tighten.

With the router motor off, set the router squarely on the template to the right of the pieces. Keep the bit clear of both work and jig, and start the motor. The first cut is made along the entire outside edge of the template, from right to left (Figure 3-15). Don't cut between the fingers of the template on this pass. You are only setting up to prevent chipping of the board. Now feed the router from left to right, moving into the template fingers.

Check to be sure that pieces are cut cleanly. If not, make a second careful pass.

Remove the pieces and check for fit. If all is well, go on to produce as many half-blind ½" dovetail joints as you need (Figures 3-16, 3-17).

If the drawer front overhangs the side of the drawer, decrease the measurement from the edge of the vertical board to the front edge of the template finger insert. If the side overhangs the front, increase that measurement. (This is why you start with scrap wood.)

If you want the partial pins to be in the same location on both sides of the drawer, rout the right side and right end of

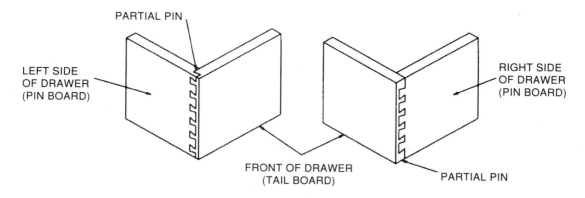

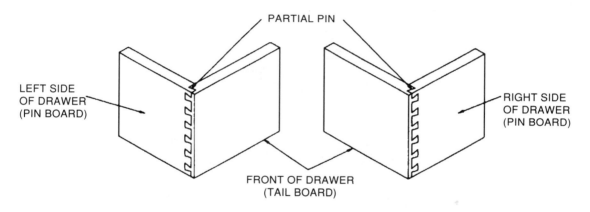

Fig. 3-16 *Arranging partial pins, from top to bottom.*

Fig. 3-17 *Arranging partial pins to be the same both sides, top or bottom*

the drawer front on the right side of the template, going through the same setup procedure, to get a mirror image.

To cut half blind ¼" dovetails, follow the same steps, with some setting changes. You need a small, accurate measuring instrument that will reach into the finger inserts of the OmniJig, giving you accurate ⅟₃₂" measurements. The best ones I've come up with are the brass insert on folding rules and the depth measuring rod on calipers. Measurements such as ¹⁹⁄₃₂" are difficult, so I pulled a brass insert from one rule and permanently marked it with red enamel. This makes small adjustments easier, too.

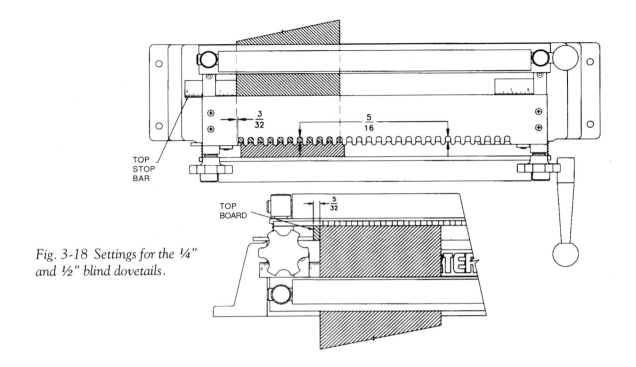

TOP
STOP
BAR

TOP
BOARD

Fig. 3-18 Settings for the ¼"
and ½" blind dovetails.

Half-blind dovetails on rabbeted door fronts work exactly as described for plain joints, but have a ⅜" wide by ⁷⁄₁₆" deep rabbet cut around the drawer front. Position the drawer front to allow for the rabbet, and cut the drawer front pins, making sure the drawer sides are in ⁹⁄₁₆" (for ½" dovetails) or ¹⁵⁄₃₂" (for ¼"). Rout the dovetails individually, instead of sweeping the entire template (Figure 3-18).

The Handmade Look

For a handmade look, use the OmniJig with a template to cut dovetails with 2" pin spacing (Figure 3-19). Tails and pins are cut separately, with two templates. There are width limits, with specific ratios looking best with this width dovetail. Multiply the number of pins times two, and add ⅛" to get the ideal width. Tolerance is plus ⅜" and minus ⅛". You thus avoid partial pins that have to be cut off before assembly.

Tail boards for the handmade look are routed first.

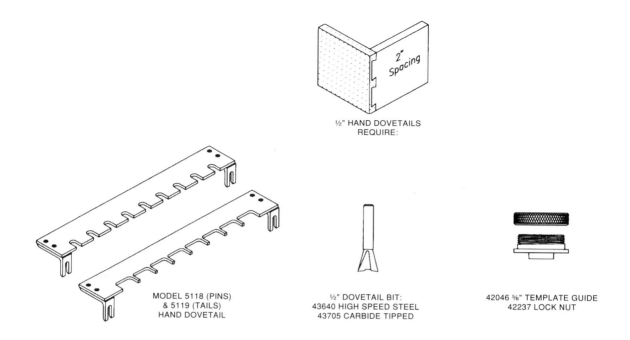

½" HAND DOVETAILS
REQUIRE:

MODEL 5118 (PINS)
& 5119 (TAILS)
HAND DOVETAIL

½" DOVETAIL BIT:
43640 HIGH SPEED STEEL
43705 CARBIDE TIPPED

42046 ⅝" TEMPLATE GUIDE
42237 LOCK NUT

Fig. 3-19 Templates, guides, bits for the wide spaced ½" dovetails.

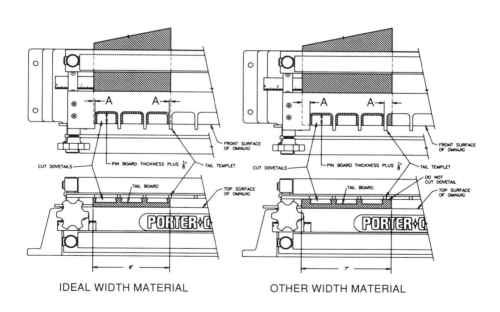

IDEAL WIDTH MATERIAL

OTHER WIDTH MATERIAL

Fig. 3-20 Setting widths, tail.

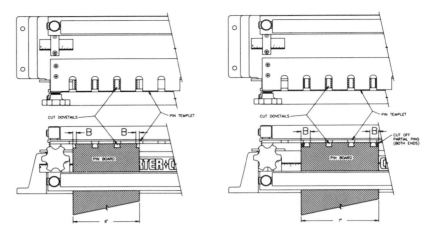

Fig. 3-21 Setting widths, pin.

IDEAL WIDTH MATERIAL   OTHER WIDTH MATERIAL

(Figure 3-21). Place in the top clamp of the jig, with a pin board clamped in the front clamp to position the tail board. The tail board is set inside up under the top clamp, butted against the pin board. Place the template so the distance from the back edge of the slots equals the thickness of the pin board, plus ⅛". This distance must be the same at both template ends. The board is then centered in the fingers and clamped. Stop bars are moved into place and locked. Routing is done as with the ½" template, with more care, for there are spots where the template guide doesn't touch the template. If the tails do not match up equally, the last partial dovetail is NOT cut. If both ends of the board are cut, rotate the board, keeping the inside up, and rout.

Rout the pin board next, with a different template (Figure 3-21). Place a scrap piece of material, about 1" wider than the tail board and the same thickness. Set its front edge as if it were the tail board. This backup board helps reduce tear-out when routing the pins. The pin template is placed on top of this scrap piece, with the front edge overlapping the pin board evenly. The pin board is placed under the front clamp, inside out, butted against the bottom of the pin template, and centered just as the tail board was. Slip the left front stop bar into place, lock, and rout the pins. Material not of a width that works with our formula above will have partial pins left. These are routed or cut off with a saw.

The opposite end of the board is routed in the same manner, after being rotated with the inside kept out (Figures 3-22, 3-23, 3-24, 3-25).

The features described make the OmniJig more versatile than other dove-tailing jigs. It is in the final three jigs that the versatility really shows. You can rout adjustable through dovetails in any pin width, and varying pin widths (on the same board end), to give a true hand-cut look. Other templates are used to cut sliding dovetails and finger joints.

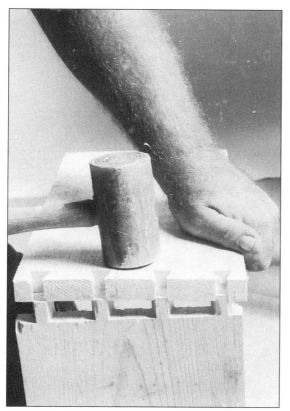

Fig. 3-22 These photos show how to assemble a wide-spaced ½" dovetail joint. A rawhide mallet is a good assembly tool. It is light, and within reason won't damage wood, so will not break the wood if the parts are not sized exactly right.

Fig. 3-23 In this routed wide part of the joint, note on the left where the bit slipped in the collet, digging into the joint a fraction of an inch.

Fig. 3-24 Pins routed for wide set joints.

Adjustable Finger Template

The adjustable finger template is different. Set the top clamp bar to secure material ¼" thicker than the boards to be dovetailed. If not provided for, your bit will hit the jig base, because all boards to be dovetailed are located *behind* the front clamp. Position thick and thin spacers on the template bracket rods to move the rods out about ⅛" from the front face of the jig (Figures 3-26, 3-26A).

Tails are cut first. Bring the front left stop bar back inside the holder so it will not interfere, then lock it in place so vibration doesn't drop it on the floor. Adjust the front clamp bar to hold the material; no need for the ¼" extra the top clamp bar needs (Figure 3-27).

Place under the top clamp a scrap wood piece ¼" thicker than, and the same width as, stock being cut. Place the template on top of the scrap piece of wood. If this scrap is narrow, slide it from side

Fig. 3-25 *Assembled wide spaced dovetail joint.*

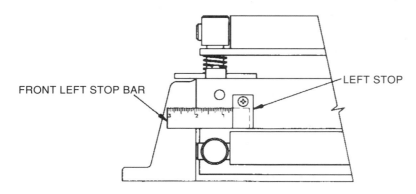

FRONT LEFT STOP BAR

LEFT STOP

Fig. 3-26 *Setting front clamp for adjustable pin template. Courtesy of Porter-Cable.*

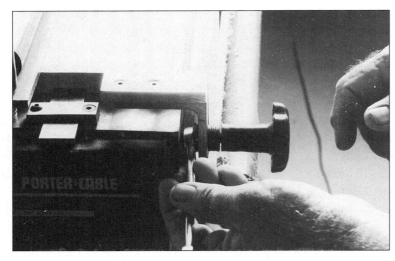

Fig. 3-26A Adjustment is needed to bring the template out at the connection to the jig.

to side to make sure the template is the same height all the way across. Place the tail board inside out under the front clamp, against the left stop, and against the template underside, then clamp (Figure 3-28).

Loosen the locking screws on the template fingers (the maker calls them forks) and slip them to their desired location. Tighten. The plastic caps on the locking screws may stick up above the template. Pry them off and reset them to cut router base interference (Figure 3-29).

Most router bases are between 6" and 7" wide, limiting spacing between fingers to not more than 3" or 3½".

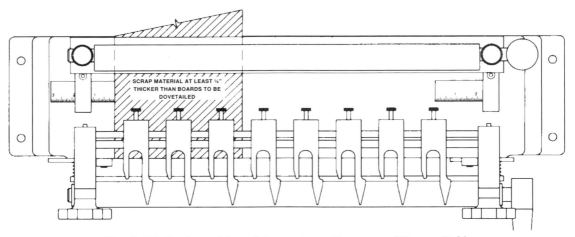

SCRAP MATERIAL AT LEAST ¼"
THICKER THAN BOARDS TO BE
DOVETAILED

Fig. 3-27 Setting tail board for cutting. Courtesy of Porter-Cable.

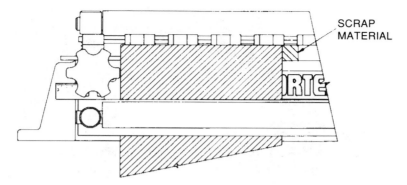

Fig. 3-28 *Setting scrap material.*

Otherwise, the router base slips in and messes up the cuts. The simplest way to get around this limitation is to increase router base width by making your own base. I once used clear ¼" polycarbonate to produce a 6" by 12" base, allowing cuts up to 6" across. Do the job with either polycarbonate or tempered hardboard, though the hardboard must be limited to about 10" in width, and won't wear as long as the polycarbonate. If extra guide fingers get in the way (this happens with wide spacing particularly), loosen them and slide them off (Figure 3-30).

This jig uses a ¾" dovetail bit, so the smallest recommended router would be about 1½ horsepower. When cutting the tails, add ¹⁄₃₂" to the depth of cut. Sand the joint flush later. Slip a scrap piece of wood against the tail board. Scrap wood backup serves to prevent tear out of wood.

Rout carefully to keep from striking the jig. If the other end of the board is to be routed, rotate top to bottom, keeping the same board face out, clamp, and rout.

To set up for the pin board, reposition the template by locating it between the thin spacer and the first black (thick) spacer. Bring the left stop out of hiding and lock it ½" over. If the material is the same thickness, clamp in the same board setting. Use the same piece of scrap stock to set the template. Tighten.

Place the pin board, with the outside facing out, butted

Fig. 3-29 Cutting.

against the left stop. The template fingers *must* remain in the same position as for the tail board.

A 5⁄16" straight bit is used for this cut. Adjust to match the thickness of the *tail* board, plus ½". Use a piece of scrap wood to prevent tear out, and rout carefully after clamping. If both ends are to be dovetailed, rotate the board as before, with the same side facing out, the lower

Fig. 3-30 Use care when spacing of pins is greater than 3"

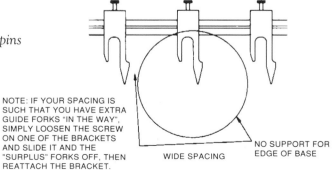

NOTE: IF YOUR SPACING IS SUCH THAT YOU HAVE EXTRA GUIDE FORKS "IN THE WAY", SIMPLY LOOSEN THE SCREW ON ONE OF THE BRACKETS AND SLIDE IT AND THE "SURPLUS" FORKS OFF, THEN REATTACH THE BRACKET.

WIDE SPACING

NO SUPPORT FOR EDGE OF BASE

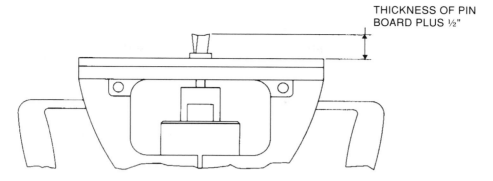

THICKNESS OF PIN
BOARD PLUS ½"

Fig. 3-31 Depth of cut adjustment.

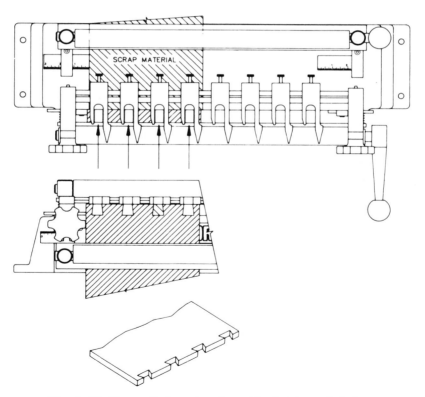

SCRAP MATERIAL

Fig. 3-32 Rout tails by using the guide slots in each fork.

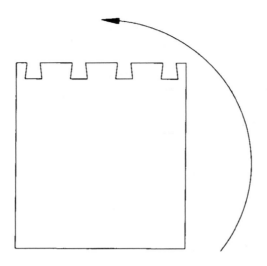

Fig. 3-33 Rotate the tail board in this manner.

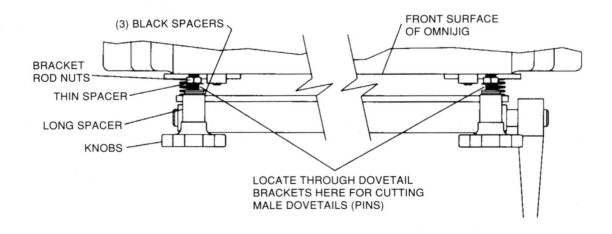

(3) BLACK SPACERS

FRONT SURFACE
OF OMNIJIG

BRACKET
ROD NUTS

THIN SPACER

LONG SPACER

KNOBS

LOCATE THROUGH DOVETAIL
BRACKETS HERE FOR CUTTING
MALE DOVETAILS (PINS)

Fig. 3-34 Reposition the template.

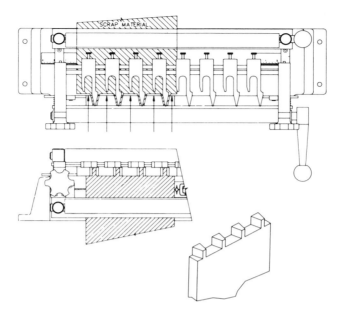

Fig. 3-35 Routing the pin board.

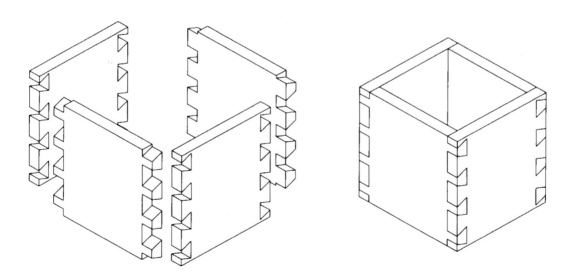

Fig. 3-36 Checking the fit.

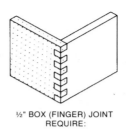

½" BOX (FINGER) JOINT
REQUIRE:

MODEL 5123
BOX JOINT TEMPLATE

½" STRAIGHT BIT:
43321 HIGH SPEED STEEL
43318 CARBIDE TIPPED

42046 ⅝" TEMPLATE GUIDE
42237 LOCK NUT

Fig. 3-37 Accessories needed for box joints. Series courtesy of Porter-Cable.

right corner coming up to become the upper left corner (Figures 3-31, 3-32, 3-33, 3-34, 3-35, 3-36).

Finger (Box) Joints

The tails and pins are not tapered, so the finger (box) joint is far easier to lay out and to cut than are dovetails. The OmniJig template allows box joints to be cut with a router, speeding up the job (Figure 3-37).

The OmniJig cuts pins and tails of the joint at the same time, reducing time spent in cutting, over and above the time saved from having to cut out each pin individually on a table saw. The short board limit is about 3', while the top board is limited in length only by available shop space and supports.

The finger joint can, with a little trimming, be turned into a hinge joint. Sand the pins so they don't lock in the tails. Drill and fit a dowel through the assembly for an effective, attractive hinge.

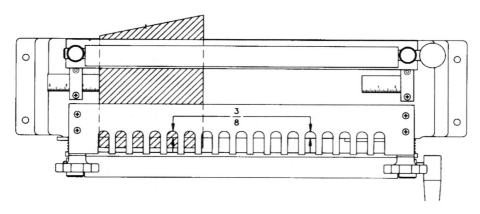

Fig. 3-38 Clamp board.

The finger joint is one of my favorites, as it meets tons of needs, is simple to make, and is attractive as well as strong.

The ½" finger joint that can be produced by the OmniJig requires material ¼" thicker than the boards to be joined in the top clamp. Again, the reason is to keep the bit from hitting the jig base. Material up to ¹³⁄₁₆" thick can be joined. The left side stop bar goes into hiding again. Scrap material is placed under the top clamp, and the template positioned ⅜" from the back of the template slot to the front face of the jig, at *both* ends. Tighten. Mark the board center and slip the board under the front clamp, against the underside of the template. Put the center mark ⅛" to the right of a template finger clamp (Figure 3-38).

Slide the left stop bar against the board and lock. Place scrap material against the top of the board to prevent tear out. Set the ½" straight bit at the thickness of the material being joined, plus ¼" (Figure 3-39).

Check to see that the bit is going to clear the jig base when it enters the scrap

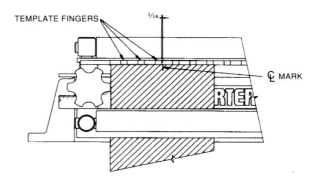

Fig. 3-39 Align template fingers.

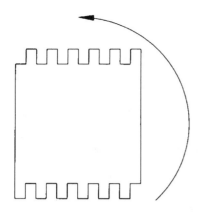

Fig. 3-40 Rotate board this way.

material. If not, use thicker material. Rout, then rotate the piece as described earlier. Butt against left stop, against template, clamp, and rout. Repeat for all sides of the box being joined, using the same settings (Figures 3-40, 3-41).

Once you've figured all your base setups, this system is faster than using a table saw to cut box joints, and just as accurate, though it limits you to a single size, where a table saw allows you to use any size you wish.

Tapered sliding dovetails are excellent for joining permanently fixed bookcase shelves to the sides of the bookcase, or for joining shelves to cabinet sides. The template looks odd, compared to other router jig templates, because it has no forward facing fingers, or forks. There are two slots extending most of the width of the template instead (Figures 3-42, 3-43).

Router setup is the same as for ½" blind dovetails, except for depth of cut, which is deeper.

We are not providing all the measurements here for any commercial jig because those measurements are subject to too many changes should the maker decide to use different suppliers or improve models. With these jigs and templates, use nail polish or enamel to place the measurements for that jig on the legs of the template brackets so they will not easily be scraped off. If the

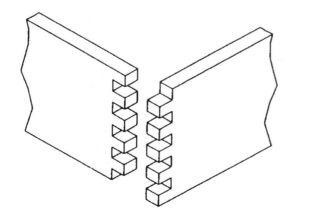
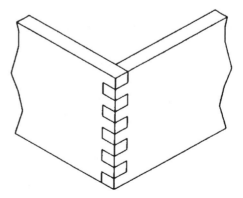

Fig. 3-41 Check fit.

Fig. 3-42 Sliding dovetail jig, in place and ready to use.

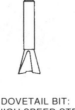

TAPERED SLIDING DOVETAIL
REQUIRE:

FOR USE WITH ¾"
MINIMUM THICK STOCK

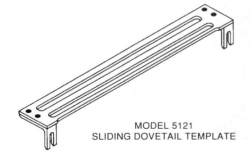

MODEL 5121
SLIDING DOVETAIL TEMPLATE

½" DOVETAIL BIT:
43640 HIGH SPEED STEEL
43705 CARBIDE TIPPED

42046 ⅝" TEMPLATE GUIDE
42237 LOCK NUT

Fig. 3-43 Parts needed to cut the sliding dovetail. Courtesy of Porter-Cable.

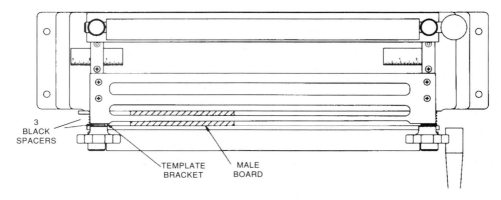

3
BLACK
SPACERS

TEMPLATE
BRACKET

MALE
BOARD

Fig. 3-44 Male board set. Courtesy of Porter-Cable.

manual is lost or destroyed, you can still work. Bobby Weaver, a good friend, stamps his measurements on the templates, eliminating any chance they'll disappear. He does the same with serial numbers of the templates, which cuts down confusion on templates that appear alike, but aren't.

Place a board the same thickness as the female dovetail board under the top clamp, flush with the front edge of the jig. Clamp. Place the template on the board, using three black spacers in front of each bracket, and a single thin spacer behind each bracket. Loosen the left front stop

Fig. 3-45 Router in place for cut, to the right of the board, from the woodworker's point of view.

bar (Figure 3-44). Place the male dovetail board under the front clamp, slipping it up to the template and as far to the left as it will go. Slide the left front stop bar to the board and lock in place, after clamping the board tightly. Adjust the template so that the male board is centered between the front two male slots on the template. With the router off, insert the bit into the template slot, to the right of the board (Figure 3-45). Move the router left, shut the router off, and let it stop completely. Repeat the process on the other slot, but move from left to right.

Before you move the male board, move the top left stop so that it will align the female (slotted part) board with the male board. Remove the male board, and slip the female board forward, with the edge against the stop. Slide a scrap piece of material under the front clamp and to the right of the board. This acts as a right side stop (Figure 3-46).

My table for this jig has a back support, but if the board extends more than 2' beyond any point, use outside support as well (Figures 3-47, 3-48).

For routers with bases larger than 5¾" in diameter, loosen the template knobs and set all spacers behind the bracket.

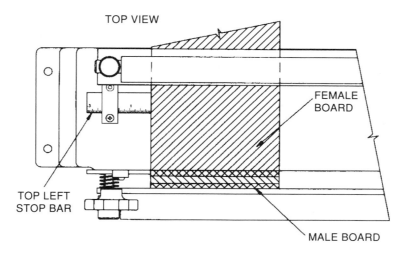

TOP VIEW

FEMALE BOARD

TOP LEFT STOP BAR

MALE BOARD

Fig. 3-46 Starting to align female board. Courtesy of Porter-Cable.

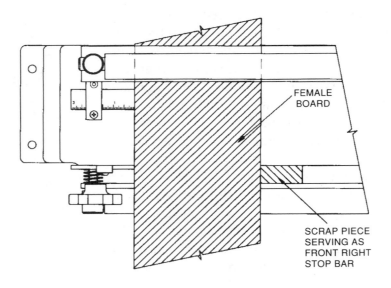

FEMALE
BOARD

SCRAP PIECE
SERVING AS
FRONT RIGHT
STOP BAR

Fig. 3-47 *Board is aligned. Courtesy of Porter-Cable.*

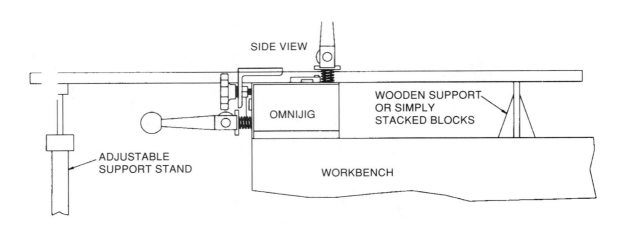

SIDE VIEW

WOODEN SUPPORT
OR SIMPLY
STACKED BLOCKS

OMNIJIG

ADJUSTABLE
SUPPORT STAND

WORKBENCH

Fig. 3-48 *Supply support to keep board from sagging. Courtesy of Porter-Cable.*

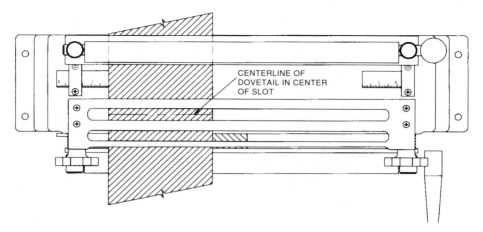

Fig. 3-49 *Mark cut along centerline.*

CENTERLINE OF DOVETAIL IN CENTER OF SLOT

Place the centerline of the required slot in the center of the template slot. In both cases, this template works best if you measure and scribe centerlines on both male and female groove ends. Set the router on the template to the left of the female board as far as you can get it. Make sure the bit is clear, and make one pass from left to right guiding on the back edge of the template slot, and another pass from right to left guiding on the front edge (Figure 3-49).

This is a simple cut to make. The only adjustment likely to be needed is for depth. If the male board tightens up before it is fully seated, the cut depth is too great, so should be reduced slightly. If the male board is too loose, depth of cut is not great enough. A bit of hand scraping can be used to seat the joint, but it is better to make adjustments in settings. If scrap wood is used to perfect your settings, you won't waste far more costly woods (Figures 3-50, 3-51, 3-52).

The Keller Dovetail System

The Keller dovetail templates offer a more restricted line of joints, but are easier to set up. Again, template cost is high, but if you'll be making many dovetails, it is worth it. The Keller units make basic through dove-

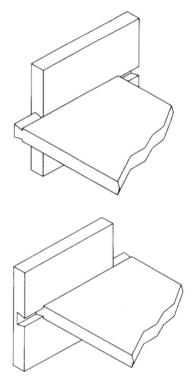

Fig. 3-50 *Check fit. If the male part slides too far in, the fit is too loose. Courtesy of Porter-Cable.*

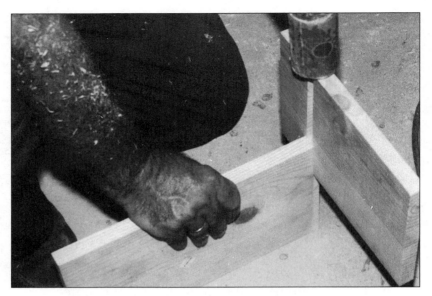

Fig. 3-51 A light drive fit is correct

tails, with fixed pin spacings, in three sizes. The templates come in pairs. One of the pair makes tails; the other produces pins. Keller includes dovetail bits of appropriate size and style with each unit. The models, with overall width indicated by the first two digits, are the 3601 (36"), 2401, and 1601. They differ in stock thickness acceptance and power requirements as well as width, the smallest being useful with lightweight routers of at least ¾ horsepower, and the others requiring at least 1½ horsepower routers. The big unit will accept stock from ⅝" to 1¼" thick, while the middle unit, probably the best buy for most woodworkers, takes stock from ⅜" thick. The small model is limited to stock ³⁄₁₆" to ⅝" thick.

The Keller system uses fixed pin units, thus are simple to set up. The use of two templates simplifies this even more, though it increases cutting time minimally. Pin spacings remain at 3" for the big unit, 1¾" on the middle unit, and 1⅛" on the small set.

You mount the templates to a backing board of your choice. Mounting holes (already drilled) are slightly oblong, and adjust until a test joint fits perfectly. I would suggest using a fir or pine backing board with the 2401 and 3601. Hardwood looks good, but because of size, weight is a problem when changing from one template to another.

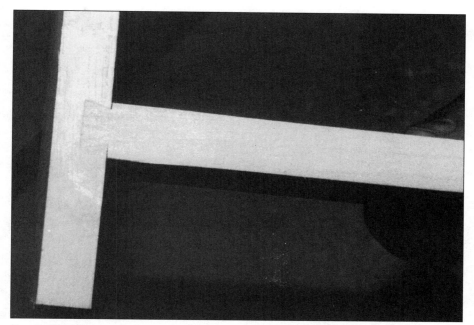

Fig. 3-52 The jig produces a superbly tight shelving or casework joint.

Fig. 3-53 Keller pin jig in use.

Fig. 3-54 Keller dovetail jig in use.

The workpiece is clamped, upright, in a bench vise. You will need a solid bench vise and bench for this work. I've mounted a Jorgenson 10" model to my assembly bench and it works beautifully.

Set the tail template on the top of the board, and center it to get your tails where you want them. Clamp the backing board to the workpiece, and rout carefully, moving from left to right. Once the tails are cut, use them to mark the pin board. Set the pin template on the marks. Clamp and rout (Figures 3-53, 3-54).

With any soft metal template, rout carefully. These units will never wear out, unless you strike the template with the router bit, which chews right through the aluminum. Keller supplies carbide-tipped bits which will never even notice the softer metal. In most cases, make sure you are engaging the template with the template guide at the start, and the bit is no longer rotating at the end, and you'll have no problems (Figures 3-55, 3-56).

I haven't tried the Leigh dovetail jigs, but their reputation is good, and utility is high with their ability to make both through and half-blind dovetails. Pin size and spacing are adjustable, and two bits, a dovetail and a straight bit, are used for the cutting. The finger template is flipped over to rout tails after the pins have been cut. Some carcass mortise-and-tenon joints are also said to be possible with these jigs, using a special template.

Router dovetailing with lower-cost jigs can provide good looking, sturdy results, without the variations of the above jigs. Usually, cheap jigs produce only half-blind dovetails, cutting pins and tails at the same time. All involve a lightweight front and top clamp, with thumbscrews or other methods of tightening. The template is on top, and the front board is brought up against the bottom of the template, while the top board is pushed against the front board, similar to the OmniJig with half-blind templates in place.

Guide bushings are used on the router, always the case when using router templates. Most of the lightweight commercial jigs will need a $\frac{7}{16}$" guide for $\frac{1}{2}$" dovetails. For $\frac{1}{4}$" dovetails, use the smaller template, a $\frac{5}{16}$" template guide bushing, and a $\frac{1}{4}$" dovetail bit (Figure 3-57).

All of these jigs are limited to stock under 1" in thickness. Most will accept widths only up to 1'. Adjustments are just as fussy as those for the OmniJig and Leigh jig. There are a few 16" wide models around.

Fig. 3-55 One template requires a dovetail bit. Courtesy of Keller Company.

Fig. 3-56 The other template needs a straight bit. Courtesy of Keller Company.

Fig. 3-57 A guide bushing set such as this covers most needs. It is made for Porter-Cable routers but may be easily adapted to others. You can buy a spare Porter-Cable 690 baseplate and mount it on your own router base.

Because the variables are so great, and there are so many different brands of these lightweight jigs on the market, it is impossible to give actual measurements for jigs beyond the above.

Some of the reasons for the popularity of the more complex and costly tools described first are simple. The lightweight jigs suffer from weak construction, wear out rapidly, and often let boards slip as they're being routed.

For a joint that can be produced with either shop-built or commercial jigs, or without jigs if your nerves and skills are both in good shape, there is the mortise and tenon.

Mortise-and-Tenon Joints

For extreme strength in applications such as legs, stretchers, and heavy holding, the mortise-and-tenon joint serves better than any other. Tenons may be single, double, or triple, as stock sizes allow and strength needs require, with mortises matching them. It is simpler, usually, to cut mortises with a spiral bit and router than it is to chisel them, but depths are limited in such cases. A ¼" solid carbide spiral router bit will reach down about 1¼". For 2½" thick material, the cut comes from both sides, a feat that may cause mismatches. Bit lengths are restricted to about 2½" in routers, with good reason. They overheat in deep work if cutting goes on too long.

Using the MorTen Device

For smaller mortises, I often work with Porter-Cable's MorTen device. This is a set of templates and a top-guided straight bit (Figure 3-58). Mortises and tenons so cut are round-edged. Care is

Fig. 3-58 MorTen bit and guide system for producing mortise-and-tenon joints.

Fig. 3-59 Haunched double tenons are just one of the many joints that may be readily produced with the MorTen device. Courtesy of Porter-Cable.

Fig. 3-60 Double tenons are quite simple with the MorTen. Courtesy of Porter-Cable.

needed when getting started with the tool. The templates are steel and touching that spinning bit to one tends to create bad nerves, if all the flying metal misses you.

The MorTen will cut mortises and matching tenons at angles, doubled, in line and in row, and provides a good joint because the templates are matched at the factory. It may, in many cases, replace doweling (Figures 3-59, 3-60).

Careful work is needed, but in essence the MorTen is set onto the work, clamped in place, and the bit is run in to cut tenons. These are cut one side at a time, after which the template is removed, flipped over, and reset on its alignment pins to allow you to cut the second side. This is the easier of the two procedures, for you can usually set the base of the router on the jig and feed the bit in. Make a careful check, first, to see that the bit clears the work before you turn on the router. If the bit doesn't clear, use a plunge router, or tilt the bit in.

Cutting mortises is a little friskier. You have to tilt the router bit into the small slot for the mortise, making sure it doesn't nick the sides or ends. Start in the center of the mortise and work outward, giving more clearance for the bit tip. A plunge router might be useful here, but only with a clear or open base through which you can view exactly where the bit is going (Figure 3-61).

To use the MorTen, clamp the wood piece into a vice, then clamp the template assembly to the workpiece. If the work is large, you may not need the vise. Clamp the MorTen to the workpiece (Figure 3-62).

Fig. 3-61 With the MorTen straight-in mortises are almost too easy. Courtesy of Porter-Cable.

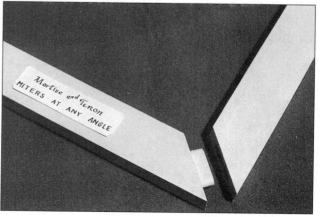

Fig. 3-62 Mitered mortise and tenons also are readily done with the MorTen. Courtesy of Porter-Cable.

Building a Jig

A shop-built jig for cutting tenons is fairly simple. Some years ago, the foreman of a shop I worked in showed me a quick jig that will serve for cutting both lap joints and tenons. It works best when you make a square offset base for your present router. Hardboard works fine here, in the same thickness as your current router base. Use a hole saw to cut a 3" hole set in 2" from one side on an 8" square. Drill the router base mounting holes so that you have a different measurement on each side. The base should extend 2" past one side of your router and flush on the other, with a ½" and 1½" extension on the other two sides, if your router has a 6" base diameter. This can be varied to suit other base sizes by adding or subtracting from the square. You then get four different cuts with one setup just by changing the side of the router against the template guide board.

The jig is made with a top of tempered hardboard or

Fig. 3-63 Drawing of tenon jig.

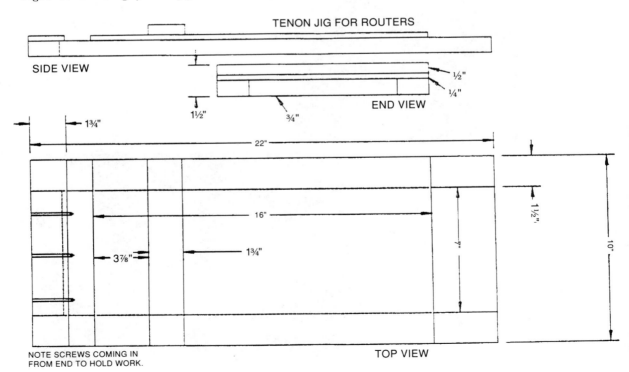

plywood, ¼" thick and 16" long. Mine is 10" wide. It is set on ¾" boards, 1½" wide, parallel to the long sides, and 22" long. The left end has a brace of the same material as the side rails, cut to fit between the side rails. There is a gap from the edge of the first sheet of hardboard to a smaller sheet covering the end rail, with the end rail sheet 1¾" wide and 10" long. A ½" by 1¾" guide board is screwed to the top of the tempered hardboard 3⅞" back from the back edge of the small piece of hardboard. Beyond that, the original model has some short, sharp nails driven in through the frame brace to hold the device securely against boards being routed. My newest model uses drive screws for that job (Figure 3-63).

The jig is butted on a square end of the board to be tenoned. To get a good seat with the nails or screws, hold the top end of the board against a wall or other solid surface and tap the jig end gently with a hammer. The router, with a straight or mortising bit of the appropriate size (as described, a ½" or ¾" bit is best) set to the correct depth, is then run across the clamped jig, guiding on the board. A change of sides on the router base and the next pass is made, until the mortise is finished on that side. The process is repeated on the other side, and, if necessary, on the ends. Make sure the board being cut butts against one of the side rails on the jig.

To cut end half-laps with this jig, butt the board, clamp, and cut to a depth half the thickness of the board. End laps do not have to be half-laps. You can cut one piece to two-thirds the board thickness, and the other to one-third, or any combination you need.

Jig for Mortises

There is a shop-built jig that works well for mortise joints. It is most easily used with a plunge router. With standard routers, some extra thought is needed as the router bit is tilted into the work. Similar in construction to a miter box, the jig is easy to make. Sizes are variable, too.

Mine started with a three-sided box, of ¾" plywood, 6" deep overall. The plywood is capped, at its top, with ¾" maple rails. The box is 4" wide inside, but your design may vary. Some routers work best with no more than 3" of inside dimension which allows the router base to rest on the sides.

The base is ¾" plywood, laminated into a 1½" thickness because this extra thickness gives the sides more rigidity. Sides must be at exactly 90° to the bottom to keep the mortise from skewing. The plywood saves money and is less prone to warp over time. Sides receive screws (#10 x 2") at intervals of about 1'. My guide is 3' long, so took four per side, all countersunk.

Guides are placed in the sides using hanger bolts or T nuts, with wing nuts to clamp the guides in place. The guides are actually end stops, and are slotted and grooved to allow adjustment under the wing nuts. These may be formed from wood or plastic.

When the stop guides are set, this jig will allow you to cut mortises with any plunge router, paying attention only to width markings. It can be set up to use width guides as well, but the extra complexity isn't worth the hassle.

As with any router-cut mortise, this guide will provide rounded corners on the mortises. Either round the tenon edges to fit the mortise or chisel out the edges of the mortise to fit the square tenon.

If you flip the stock, this jig gives a maximum depth of close to 5", depending on the length of the router bit and the size of your router. To brace stock at the proper height, insert flat stock under the stock to be mortised until it reaches the depth needed with the bit you have selected. Clamp and rout.

To reduce problems with heat buildup and dulling of bits, use solid carbide spiral bits, not high-speed steel (HSS) straight-faced bits. Spiral bits lift chips from the holes more efficiently than do straight-faced bits, which dull rapidly and heat up from the waste packed around them. Use a ¼" bit for mortises up to ½" wide, and ½" bits for mortises ½" and wider (Figure 3-64).

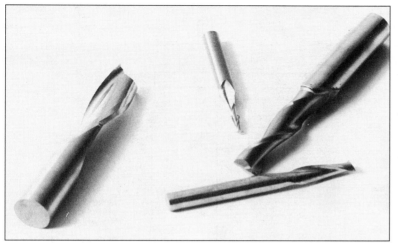

Fig. 3-64 Spiral routing bits, these from Trend-Line, come in a variety of sizes and widths and work much better than straight bits in deep routing.

Other Joints

Routers can also make joints for projects that do not require expensive jigs, nor much in the way of shop-built equipment beyond a router table, if that. Router tables may be bought or built, but building one is not particularly difficult, as you will note from the plans shown here for mine (Figures 3-65, 3-66, and 3-67).

Some of the same bits that are useful on a router table may also be used, with great care, freehand, but others are not at all suitable for freehand use. Freud's box joint bit is a case in point. It requires a powerful router to work well, and has a strong tendency to throw the work, or jerk the router around. It is far safer to mount the bit in a table (Figure 3-68).

Other router bits may be used freehand, but do a smoother job on a router table. All safety rules, regardless of bit used, need to be observed. Larger bits require more care, as do more powerful routers. There is an added check to make when using any large bit with a plunge router: it is essential that the bit not spread wider than the opening in the metal base of the router. Should it have such a spread, do NOT use it with that router, as retraction of the bit will cause it to strike the base of the router. You can make bases with larger openings for most routers.

Fig. 3-65 Router table plan.

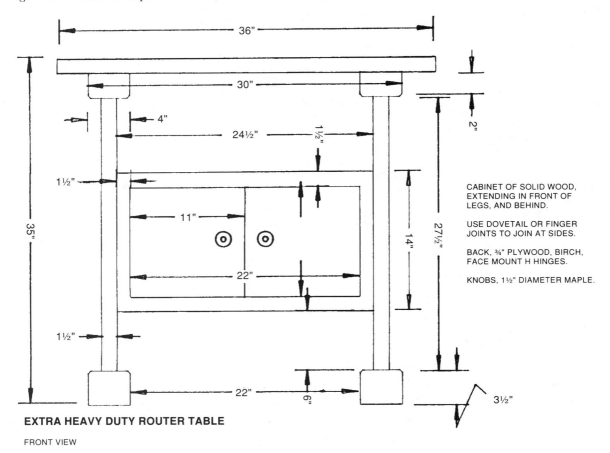

CABINET OF SOLID WOOD,
EXTENDING IN FRONT OF
LEGS, AND BEHIND.

USE DOVETAIL OR FINGER
JOINTS TO JOIN AT SIDES.

BACK, ¾" PLYWOOD, BIRCH,
FACE MOUNT H HINGES.

KNOBS, 1½" DIAMETER MAPLE.

EXTRA HEAVY DUTY ROUTER TABLE

FRONT VIEW

Fig. 3-66 Router table plan (cont.)

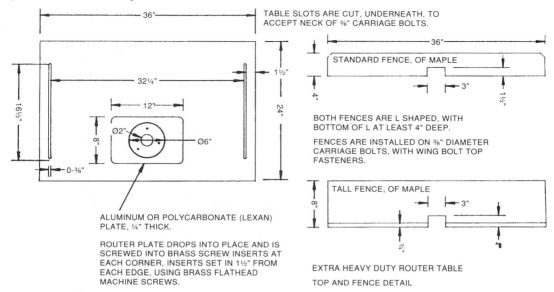

TABLE SLOTS ARE CUT, UNDERNEATH, TO ACCEPT NECK OF ⅜" CARRIAGE BOLTS.

STANDARD FENCE, OF MAPLE

BOTH FENCES ARE L SHAPED, WITH BOTTOM OF L AT LEAST 4" DEEP.

FENCES ARE INSTALLED ON ⅜" DIAMETER CARRIAGE BOLTS, WITH WING BOLT TOP FASTENERS.

TALL FENCE, OF MAPLE

ALUMINUM OR POLYCARBONATE (LEXAN) PLATE, ¼" THICK.

ROUTER PLATE DROPS INTO PLACE AND IS SCREWED INTO BRASS SCREW INSERTS AT EACH CORNER, INSERTS SET IN 1½" FROM EACH EDGE, USING BRASS FLATHEAD MACHINE SCREWS.

EXTRA HEAVY DUTY ROUTER TABLE

TOP AND FENCE DETAIL

Fig. 3-67 Router table plan (cont.)

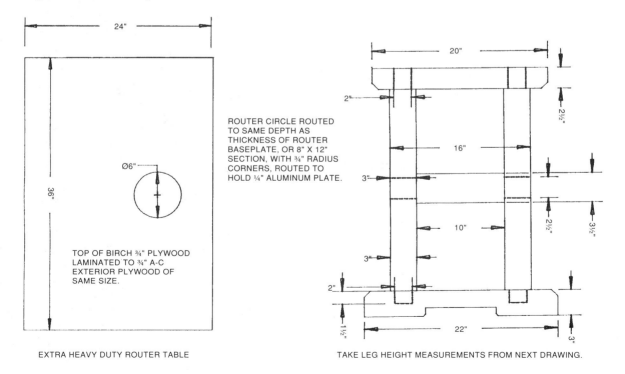

ROUTER CIRCLE ROUTED TO SAME DEPTH AS THICKNESS OF ROUTER BASEPLATE, OR 8" X 12" SECTION, WITH ¾" RADIUS CORNERS, ROUTED TO HOLD ¼" ALUMINUM PLATE.

TOP OF BIRCH ¾" PLYWOOD LAMINATED TO ¾" A-C EXTERIOR PLYWOOD OF SAME SIZE.

EXTRA HEAVY DUTY ROUTER TABLE

TAKE LEG HEIGHT MEASUREMENTS FROM NEXT DRAWING.

Fig. 3-68 Freud's heavy box joint bit.

Splined Joints

Splined joints are another area where routers shine. Splines may be applied to miters or butt joints. The router is used to slot the edges to be joined, using a slotting cutter. If you are doing miters, the easiest way to stay with the angle is to clamp two mitered pieces together so they form a point at the top. Use a good bench vise as a base for your clamping and add clamps as needed for longer miters. Insert the slotting bit in the router and adjust for height, with the router base on the opposite side of the point. Do setups with the router unplugged. Make a test cut in scrap stock, then rout the spline grooves from each side of the point with the router base riding on the opposite side. The router rests on one side of the point; the bit cuts into the other.

For simple butt joint splines, clamp and set the depth. Then rout the groove with the pilot guide riding on the board edge.

End laps are easily made as described earlier, but center laps are a bit different. A straight or mortising bit is used, with the router riding along a clamped-on straightedge to give the correct width to the lap. Depth is carefully set in same-size scrap stock before final cutting.

Cutting Dadoes

Dadoes are quickly and readily cut with routers. There are jigs you can use to make sure the spacing is correct. The simplest is a set of spaced bars. The router moves inside them. The space between bars needs to equal the width of the router base. Under the side guide bars are

Fig. 3-69 *Dado jig drawing.*

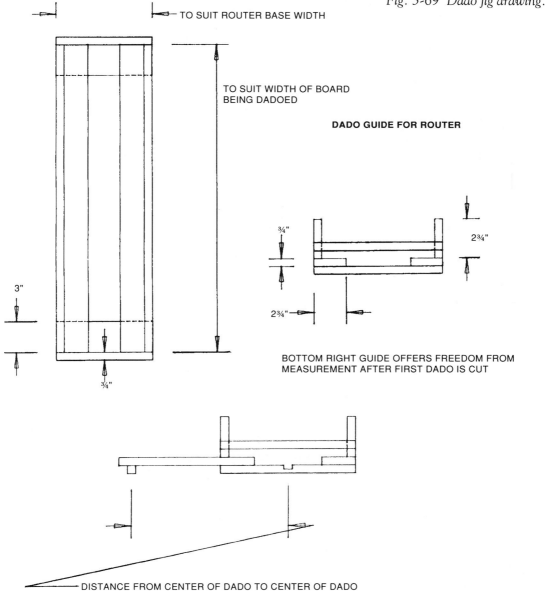

TO SUIT ROUTER BASE WIDTH

TO SUIT WIDTH OF BOARD BEING DADOED

DADO GUIDE FOR ROUTER

3"

¾"

¾"

2¾"

2¾"

BOTTOM RIGHT GUIDE OFFERS FREEDOM FROM MEASUREMENT AFTER FIRST DADO IS CUT

DISTANCE FROM CENTER OF DADO TO CENTER OF DADO

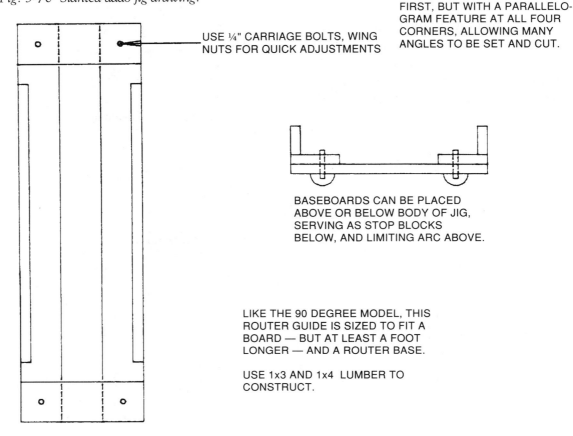

Fig. 3-70 Slanted dado jig drawing.

THIS GUIDE IS SIMILAR TO THE FIRST, BUT WITH A PARALLELO-GRAM FEATURE AT ALL FOUR CORNERS, ALLOWING MANY ANGLES TO BE SET AND CUT.

USE ¼" CARRIAGE BOLTS, WING NUTS FOR QUICK ADJUSTMENTS

BASEBOARDS CAN BE PLACED ABOVE OR BELOW BODY OF JIG, SERVING AS STOP BLOCKS BELOW, AND LIMITING ARC ABOVE.

LIKE THE 90 DEGREE MODEL, THIS ROUTER GUIDE IS SIZED TO FIT A BOARD — BUT AT LEAST A FOOT LONGER — AND A ROUTER BASE.

USE 1x3 AND 1x4 LUMBER TO CONSTRUCT.

end bars to equal the width of the router base plus the width of the guide bars. Use straight 1x3 or 1x4 stock for all pieces. Make sure the corners are square. Screw and glue assembly assures long-term use (Figures 3-69, 3-70).

Butt the jig on the side of the board to be dadoed. Clamp lightly on the end away from where you'll start cutting. The first cut will produce a cut in the jig's end spacer; mark the center of that cut width. The mark makes line-up for succeeding dadoes easy. Set depth carefully. The router is started and run across the piece being cut.

Guide bars are longer than the stock being cut. Just because you are now routing a single 12" wide piece, don't figure that will be the widest you'll ever cut. It is often easier to match pieces — for bookcase sides and other projects — if all cuts are made on all pieces at the same time. Making the side bars at least 3' long is a good idea.

For stopped dadoes, place stop blocks on the guide bars, clamping them across where needed. Stopped dadoes done with a router will have rounded ends, giving the choice of rounding the piece to be inserted or squaring off the rounded dado ends with a chisel.

This guide presents a number of ways, inexpensive and expensive, easy and finicky, to rout various joints. The more difficult the joint to make by hand, the more difficult it is to set up to make with any power tools.

Working for ourselves, we want to make a choice between cost and convenience, and will have, at most, short production runs for any particular setup. The more time spent setting up, though, the easier and more attractive the final job will be. Machines can help.

CHAPTER 4

MAKING JOINTS WITH POWER SAWS

Power saws provide great versatility in cutting joints, with the edge going to the table saw, a tool as versatile in making joints as the router.

Table saws come in a wide range of types, sizes, and styles, some light enough to be moved from one job site to another, and others not readily moved at all. My basic table saw does not move 6" from year to year. It weighs upwards of 600 pounds. The saw itself weighs 400 pounds, while the Excalibur rip fence and Excalibur sliding table add the rest of the weight and increase the usefulness of the tool.

Selecting a table saw is not a simple job because of the variety on the market. Blade sizes range from 4" on up to 14", while power goes from fractional horsepower to three-phase multi-horsepower. It is a rare home shop that needs three-phase electrical power. That is primarily for industrial needs, and is found mostly in 5 horsepower and up tools (Figure 4-1).

Once you decide on a 1 to 3 horsepower motor, think about the quality of the rest of the saw. Tolerances need to be close, the table precisely machined, and the adjusters properly made and installed. The miter slot must be precisely cut, and the rip fence well made. Both of these may be replaced with accessories such as the Excalibur units, but remember that such accessories cost hundreds of dollars. They do expand the capabilities of any table saw, but also limit its portability. (My Excalibur fence and sliding table — neither of them the largest models available — make my Unisaw something over 8½' wide, as well as heavy.)

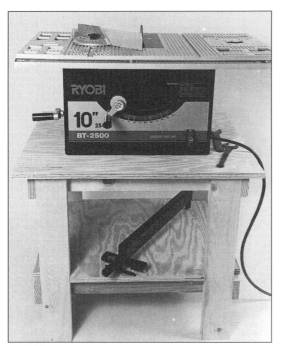

Fig. 4-1 Ryobi's 10-inch table saw is a good example of the accurate lightweight models.

Consider Your Uses

Think how you will use a table saw. The heavier and more powerful the saw, the less portable but the more accurate it will be. Contractor's saws are powerful and accurate, but barely portable; light production saws are more accurate, and not portable.

If dead-on accuracy is of major importance but portability is still a factor, consider buying a second saw. There are lightweight 8" and 10" table saws that can almost be tucked under one's arm and carted off. Several of these offer surprising accuracy from their light aluminum castings.

Next, decide which joints you will be building. Many joints are made with such difficulty on the table saw as to be best forgotten about or treated as novelties. Among such joints are dovetails. They take so much time and are so difficult to set up to produce that hand cutting is preferable.

The finger joint is a winner as a substitute for the dovetail. It does not resist pull without glue, the way a dovetail does, but does supply a huge amount of gluing surface, which is preferable with modern glues. Finger, or box, joints can be doweled through pins and tails with a single dowel (or a double dowel on really large sizes), after which it will resist pull forces without gluing. A single dowel may be used as a pivot point, the ends of the fingers rounded over, and the entire joint becomes a hinge.

Using Jigs for Finger Joints

Finger joints are a specialty of table saws, and were an American invention in the latter half of the nineteenth century or the early part of the twentieth. Several jig styles work well, one of which is the commercial Accu-Joint jig. I use an Accu-Joint, with only two objections: the tiny locating pins are exceptionally easy to lose and the jig itself is only a foot wide, limiting the width of any pieces being joined (Figures 4-2, 4-3).

That limit is not serious, because 90 percent of joints for drawers and small boxes are in stock narrower than 12". With the width limit comes the ability to easily and quickly cut finger joints in three sizes: ¼", ⅜" and ½", all readily made by laying a template over the basic template and changing the dado blade width.

Making a finger joint jig is also easy, requiring a few pieces of wood, accuracy, and a dado blade set. The dado set is needed for the Accu-Joint and other table saw work anyway.

Shop-made jigs take different size cutouts for each size of finger joint. You need a separate jig for each joint size, but you are not limited to three or four or five joint sizes.

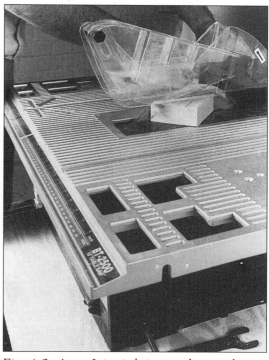

Fig. 4-2 Accu-Joint is being used to produce a finger joint.

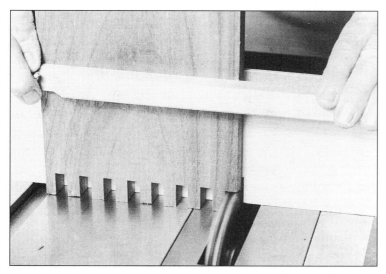

Fig. 4-3 This finger joint was made with the Accu-Joint and a Freud dado set.

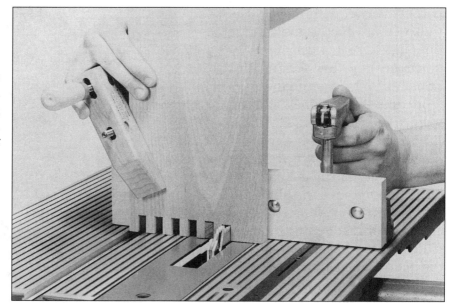

Fig. 4-4 The homemade finger joint jig, in operation, depends greatly on the accuracy of construction. Courtesy of Shopsmith, Inc.

If you need 1" finger joints, a jig can be made to produce them (though most dado heads will require two passes per joint cutout with such wide fingers). The maximum cut with many dado heads is ¹³⁄₁₆", good for single pass finger joints up to ¾" wide. More modest finger joints give a better appearance, with ½" being the largest size used for wood up to 1½" thick. The thinner the stock, the lighter the finger joint that gives top appearance.

The simplest jig is usually the best. Start with a miter gauge extension that passes the blade, or dado head, on the side you prefer to work. Measuring care is imperative with any jig. Sloppy jig measurement and assembly means joints won't fit (Figure 4-4). Use oak, maple, ash, or a similar hardwood. Softwoods may be used, but do not retain accuracy for long. The extension must be at least 3" high and 16" wide, with both height and width increasing as the size of boards to be joined are increased. The extension is securely screwed to the miter gauge, and set so the miter gauge moves freely in its slot. Don't get the extension so tight to the table that the gauge can't slide with it mounted.

Make a pass over a ⅜" dado blade, set ¾" deep. Check

dimensions by measuring, then make a practice cut in scrap stock first. Measure over ⅜" from the first slot and cut a second ⅜" x ¾" slot.

Into the first slot, insert a ¾" x ⅜" x 2½" stop block. Secure the entire jig to the miter gauge, making sure that the second, or open, jig slot is directly over the set dado blade.

Make a guide strip the width of the dado blade setting. Use that to offset the board being cut. The guide strip is placed alongside the board edge. Both are held upright, butted against the stop block. Spring clamps work well here. Making the first pass over the dado blade gives a cut on the board. Remove the guide strip, move the board over so the cut fits over the stop block, and place the mating piece in front of the already cut piece, its flat edge butted against the stop block. Clamp tightly, and pass through the dado blade. Move the entire assembly over (notch moves onto stop block) and repeat the cut. Continue until the entire board is ready. Pieces will then mate, and you have cut two sides at one time, a solid time saver over other finger joint jigs.

Fig. 4-5 Jig construction details are not difficult, but need to be precise, as does fastening of the table saw to the miter gauge. Courtesy of Shopsmith, Inc.

Need Separate Jigs

For each joint size, you need to make a separate jig. The four most popular are ¼", ⅜", ½", and ⅝". Where different sizes are needed, replace the ⅜" in the jig design with the above fractions. For the larger two sizes, it is practical to increase the length (the ¾" measurement) to 1".

For ease of fit, add 1/32" to cut depth. This allows fingers to extend far enough to allow some light sanding to finish the joint. If depth is too shallow, the only way to finish the joint is to sand the entire side.

It is possible to build a jig that will give size alternatives, not requiring a jig for each size. It is difficult to set up the jig, though it is not hard to make, and only rarely is it of real use. Great care in measuring and checking is needed for this jig to be any use at all (Figure 4-5).

Size limits for table saw jigs that make finger joints are not as restrictive as for router jigs with dovetails. The widest dovetail jig is 3' across (the Keller jig can be set and used for wider boards, of course), and expensive, while the closer to standard 16" wide Omni-Jig is also costly. Cheaper, narrower dovetail jigs are available from Porter-Cable, Black & Decker, Sears, and others. Finger joints are limited in width by the working space available and by the support available for the wood. Only the Accu-Joint jig provides a built-in width limit.

Limitations in height are another story. Dovetailing is often done on a flat board, laid flat, so while width is limited by jig width, length is limited by shop length and work supports, as width is with finger joints. Finger joints, though, are commonly cut on table saws, which take most of 3' out of floor-to-ceiling distance, leaving a 5' to 7' length possibility. Boards get very hard to handle once they get much over 48" long. Jig modifications must be made to help deal with widths over 4'.

Various ways of solving this problem are available, but it must be remembered that most jigs will twitch a bit with long boards in them, and barring massive money infusions, that twitch is not totally controllable at the saw table after a certain point. My experience has been that at 6' I'd

Fig. 4-6 Using a second board at the end not being cut helps keep things precise when the AccuJoint is used on narrow stock.

Fig. 4-7 Wider boards need no help from leveling stock.

Fig. 4-7A This box was cut with Accu-Joint and Nicholson dado set.

rather use some other form of joint, no matter how heavy the jig. Jig modifications usually will be higher miter gauge extension boards, allowing more clamping surface. If your table saw has a very heavy miter gauge and groove, this may work well; if not, it won't (Figures 4-6, 4-7, 4-7A).

GROOVING AND DADOING

Table saws do good work with dado joints and grooves. Grooves and dadoes differ. Grooves go with the grain of the wood, while dadoes go against, or across, the grain of the wood. Dado sets are simple accessories, most being stacked blade assemblies that give a wider kerf, or cut. The outer blades are similar to standard saw blades, while the inner, or chipper, blades commonly have only two teeth, at opposing sides of the blade. Chipper blades clean out the area between the outer blades, producing a set width groove, or dado (Figures 4-8, 4-9).

Most standard dado sets will accept enough chipper inside blades to give a dado width of $^{13}/_{16}$" in a single pass. Many accept more chipper blades, if your saw arbor will handle the extra width (Figure 4-10).

Single or double blade "wobbler" units are available.

Fig. 4-8 Deep groove being formed with Nicholson dado head on a Craftsman table saw.

The one or two blades in the tool are set to wobble at a maximum specified distance, producing a cut of the same size as the wobble. Such dado sets leave more material in the bottoms of the grooves, but are faster and easier to set up. (Not always: the better the wobbler quality, the easier it is to set up. Cheap wobblers often never set up properly or won't stay set up.)

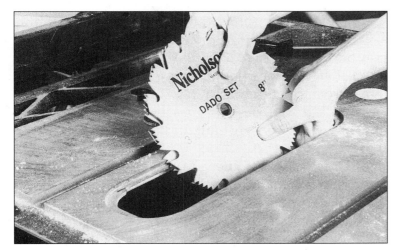

Fig. 4-9 Multiple blade dado set being installed.

Fig. 4-10 Dado set in motion.

Groove and dado joints are useful for setting in shelves in cabinetry, and in bookcases where shelves and dividers do not need to be adjustable for height. They are stronger than simple butt joints, and neater looking (Figure 4-11).

Making stopped dado and groove joints is possible, using stop blocks mounted on the saw to keep the cut to a length. Using stop blocks on such operations on a table saw needs a good bit of forethought, because the saw may have to be turned off as the stop block is reached — if the material does not allow a good, safe grip for removal from the cutting area. It is sometimes easier to use stop blocks on dadoes and grooves with a less powerful tool, one with less penchant for kicking back, such as a router.

If material is long enough — that is, if it extends far enough away from the blade — you can use stop blocks on the table saw, with a dado setup. Using such a block leaves an arced bottom in the cut. This needs to be cut away with a chisel, or the board going into the dado arced to fit the dado.

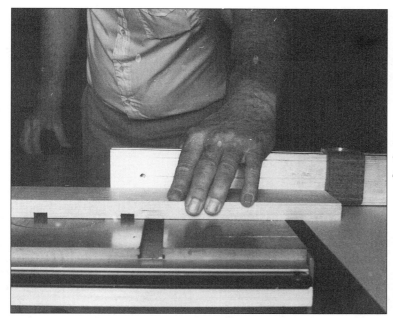

Fig. 4-11 Cutting multiple dadoes works well for shelf side stock, and gives good strength.

If you use a router to make the stopped dado, it will leave curved corners, instead of an arc rising from the bottom. Those can be chiseled out, usually with a corner chisel.

Rabbet Cuts

Dado blades are also useful for rabbet cuts, simple shelves cut into the wood, into which another piece of wood or other material fits, making a rabbet joint (Figure 4-12). A rabbet cut is set up to cut the rip fence, leaving no lip on the cut.

I prefer to work with an auxiliary fence for rabbets, reducing the chance of pinging my expensive Excalibur aluminum fence with the carbide tip of the dado blade. Use a straight piece of wood (my preference is for maple), and attach with screws through the holes in the fence. If there are no holes, drill two. Get the edge of the auxiliary fence down on the table, but not so tight the fence won't move with it in place. Now, cut a relief arc in the fence facing. This arc will need to be different depths for different cuts, so start with a slight cut at two-thirds the depth

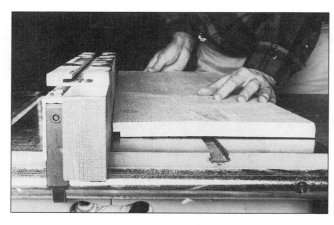

Fig. 4-12 Rabbet cuts are often used to accept back boards in cabinets and shelves.

Fig. 4-13 This Boyne gauge, advertised for router bit depth setting, also works for standard saw blades, and here, for a dado head.

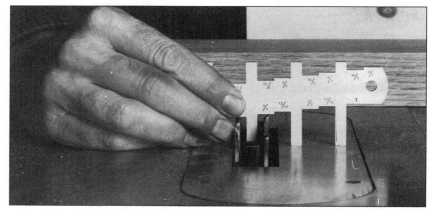

of the facing width. If you used a ¾" thick board, cut in ½", and so on. Raise the blade slowly to increase the depth of the relief arc, to a maximum height of 1".

Rabbets may also be cut with conventional saw blades, in two passes. With rabbets up to ¾", a dado blade needs only a single pass to produce a rabbet, while conventional blades take one pass to cut depth, then another to cut waste loose. That requires two blade setups as well, adding to the chances of something going wrong.

Using the dado setup to cut a rabbet, set the depth of the blades, after installing blades to give a width of about ¹⁄₁₆ to ⅛" more than the needed width (Figure 4-13). Set the auxiliary fence so that it gives the needed width, and lock all settings. Feed the material through, keeping a snug fit against the fence facing. Use push tools for any cut where the blade is going blind, which is the case with all rabbets and dadoes.

Tongue and Groove

Dado blades also serve to cut tenons, and tongues and grooves. The simplest of the cuts is the tongue and groove. Set the fence to center to proper width for cutting the groove — the distance that will be cut away on each side to form the tongue. The depth is fractionally larger than for the tongue, but no more than $\frac{1}{16}$" no matter how large the tongue dimensions are. Make the dado width somewhat smaller than the actual requirement for the groove to be cut (Figures 4-14, 4-14A, 4-15, 4-16).

Make one pass with the board, and reverse and make a second pass along the other side. The groove will be exactly centered if cut like this.

A single setup is made for the tongue. Set the dado blades to depth and width, and pass the board over so that one side is cut away. Reverse the board and cut the other side away, to finish the tongue. Do this for all boards to be tongued, but only after checking exact fit with the grooved boards. If the fit is good, make all the cuts needed. If the fit

Fig. 4-14 Often, table saw inserts will not accept the accessory blades you wish to run. My Delta would not take an 8" dado head at full height, so I epoxied together two $\frac{1}{4}$" sections of polycarbonate, bandsawed and sanded the results to fit, and taped them in place. Then I ran the dado setup through the special insert, starting very slowly.

Fig. 4-14A The epoxied together polycarbonate first has a lift hole drilled.

Fig. 4-15 The dado comes out at full height, though not full width at this time. It is important not to pop the insert loose from the tape, so take the cuts slowly.

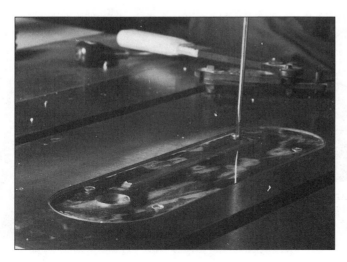

Fig. 4-16 For the Delta, drill the insert for brass inserts and screws which are used as adjusters at the corners. Others required screws to hold them in place.

is off, make any needed adjustments. Cut the tongues last because they are more easily adjusted than are the grooves.

Cutting Tenons

Tenons can be cut with dado blades. This cut is made like cutting tongues since a tongue is, in essence, a long tenon. Run the dado blade up to the height to be cut, set the width on the inside, and pass the work through. This needs some form of carriage if the work is tall. You may wish to make your own carriage, or buy one.

Dado blades can extend only about 2" above the table insert, so there are limits on the length of tenons cut in the upright position. Longer tenons are cut in a series of passes, with the dado set to the depth needed to clear material from one side of the board. Clamp the board in the miter gauge, and use a stop block on the rip fence. Make repeat cuts until the last cut comes off the stop block. You may also start with the cut from the stop block, but don't use that as one of the intermediate cuts. Flip the board over and repeat the process to cut the other side of the tenon (Figures 4-17, 4-18, 4-19).

True tenons have the edges cut away. This needs another saw setup.

Stud tenons do not have the edges cut away, and are often used as parts of what the British call bridle joints — that is, the mortise has only two solid sides, similar to a horse's bridle, so insertion is quite easy. Strength is lower, but the entire joint is easily cut on the table saw (Figure 4-20).

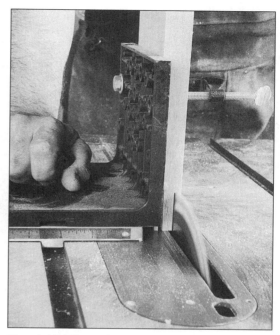

Fig. 4-17 This is the Sears' Craftsman version of a tenon jig.

Fig. 4-18 This slightly newer version of the tenoning jig is available from Trend-Lines and Woodcraft.

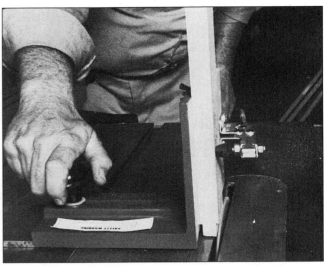

Fig. 4-19 *The jig works in the same manner as the Craftsman, and most others, but uses a newer type of clamp to hold stock in place.*

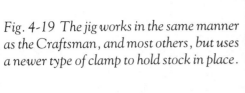

Fig. 4-20 *The tenon produced is clean.*

Splined Joints

You can cut splined joints with the dado blade and cut the slots for the thin splines with a standard saw blade. The spline may run the length of the board, or be stopped (a blind spline joint). The plate joiner is rapidly replacing various types of splining.

Standard grooves for splines are made exactly as are grooves for tongue and groove work, but are usually not cut from thick stock (most splines are thinner than ¼"). Splines are made of plywood, or of stock with grain running against the stress direction, regardless of how board grain runs. The spline grain runs 90° to the spline groove (Figure 4-21A).

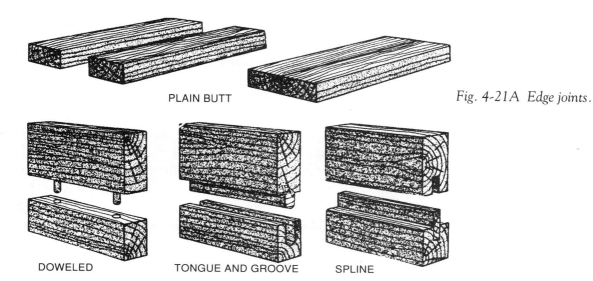

PLAIN BUTT

DOWELED   TONGUE AND GROOVE   SPLINE

Fig. 4-21A Edge joints.

Blind splines are cut with two stop blocks, one at the front of the fence and one at the back. Working at the front of the table saw, clamp stop blocks firmly to the rip fence, and brace work against the front stop. Lower onto the turning cutter, whether that cutter is a dado head or a single saw blade. Move it towards the rear stop block after it is flat on the saw table.

Form the spline to fit the groove, not the other way around. The ends of the groove will be arced, which means the splines are cut to fit that arc. It is far easier to fit the splines than it is to fit the grooves.

Notching

Additional table saw joints are possible, using a combination of standard saw blades, dado heads, and care.

Notching is a joinery method used

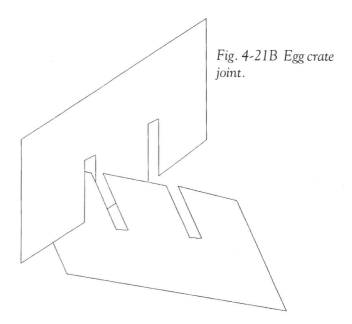

Fig. 4-21B Egg crate joint.

to make finger (box) joints and egg crate joints (Figure 4-21B). Notching is using a dado blade set to cut the appropriate width from stock, with the stock held so the cut is made across its thickness instead of its width.

For egg crate cuts, make sure the notch is the exact width of the stock thickness used, and the depth of cut exactly half the height of the stock used. You may, if you wish, mark a single piece of wood, and clamp all those needed together, thus cutting the entire project as a single unit. With the marked piece at the front, and the dado set properly, make the cuts. Turn one set over to fit into the other set. This only works with square projects. Others need two packs made up to be cut (Figure 4-22).

Lap Joints

Lap joints are rapidly made by notching, using a dado head to cut away stock. This is important for lap joints that are not at board ends, where a standard blade takes

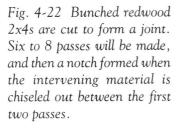
Fig. 4-22 Bunched redwood 2x4s are cut to form a joint. Six to 8 passes will be made, and then a notch formed when the intervening material is chiseled out between the first two passes.

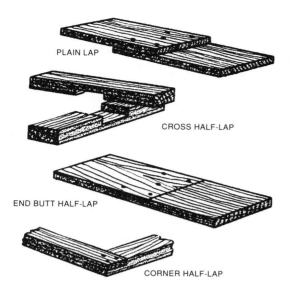

Fig. 4-23 Lap joints.

PLAIN LAP

CROSS HALF-LAP

END BUTT HALF-LAP

CORNER HALF-LAP

too long to make all the passes needed to clear the notch for the lap. The dado blade is used to set the end markings, cutting first at one end and then at the other. After that, multiple passes clear the center of the notch. The process may be repeated if both boards are to be center lapped.

End laps work in much the same manner but only the inside end is cleared first, to the marks, with multiple passes until the end of the board is reached.

Laps may be mitered, and made to match different board thicknesses, as well as being cut to half depth of two boards of equal thickness (Figure 4-23).

Miter and Butt Joints

The miter joint is a variant of the butt joint in that pieces cut at 45° angles are butted together and fastened in place. It is useful for picture frames, door and window molding, and similar objects. Miter joints are also used when working with plywood to leave only the surface of the plywood showing, covering underlying plies. The miter joint is a cabinetry joint for both moldings and frame carcass construction (Figure 4-24).

The table saw is not perfect for creating smooth

Fig. 4-24 Miter joints.

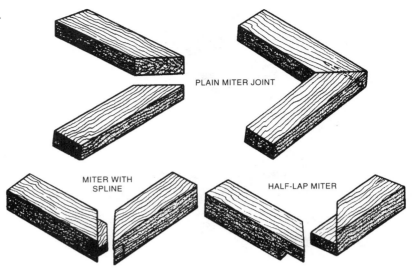

PLAIN MITER JOINT

MITER WITH SPLINE

HALF-LAP MITER

crosscuts. The radial arm saw is better, but is limited in crosscut capacity. My DeWalt 12" radial arm saw is large for a home shop saw, but has a crosscut capacity, in ¾" material, of only 16". It will cut only 14" in thicker material. To get wide crosscuts, it's necessary to go to industrial radial arm saws such as Delta's 18" (you will not like the overall tool or blade prices on this one). Even then, the crosscut capacity is 23¼", just about what is needed to slice across a kitchen counter top.

Fig. 4-25 Ryobi's small radial arm saw cross cuts and miters well.

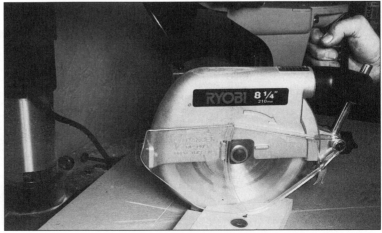

To miter a counter top, you need larger tools with a capacity of 19" left, 11" right. Ryobi's folding radial arm saw offers a decent crosscut — ¹⁄₁₆" short of 11" — but not a wild one, though on a cost comparison, it is better than the larger saws and costs about a quarter of what the DeWalt model costs (Figure 4-25).

Build a Miter Box

For most common widths of crosscuts, the radial arm saw tends to be superior to the table saw, unless you find one of the sliding tables now made for table saws. You might also create a table saw miter box yourself, easing the chore of getting accurate crosscuts and miters. I use a 34" sliding table now, (and wish I had known a 50" was available when I got this one), but the sliding miter box is the most sensible for owners on a budget.

Start with two pieces of hardwood cut to fit the miter gauge slots. Your saw may have different size slots, but most are around ⅜" x ¾". Use hardwood only, and make the length at least 2'. Maple is preferable to oak. Place these in the miter gauge slots, evening them up with the front of the table (saw blade lowered all the way). Make sure the fit is good and that the guides slide readily, without looseness. Coat the tops with glue, and clamp your jig

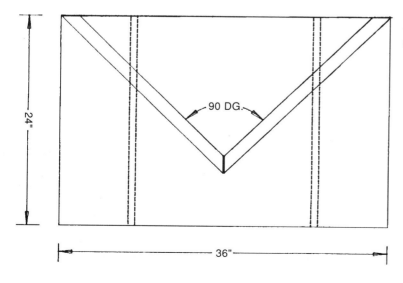

Fig. 4-26 Miter jig for table saws.

90 DG.

24"

36"

table top, even with the front of the saw table, in place. Jig top size depends on table saw size, but a minimum of 24" deep (measurement parallel to the saw blade) and 24" wide is recommended. If the table is larger, wider is also better, but extended miter gauge slot runners and bridging may be added to make very wide miters, if you have a saw capable of handling such cuts (Figure 4-26).

Allow the adhesive to set, and remove the jig. Drill and countersink three holes per guide strip (one 2" in from each end and one centered). Place the screws.

Slide the table back and raise the blade. Make a partial cut through the jig table top at a 90° angle, straight on into the blade. Leave at least 6" of material at the back end.

Mark lines at 45° angles on both sides of that cut line, and cut and place a 2x3 (planed flat: do NOT use standard 2x3 studs; plane a 2x4 to size to get sharp corners) on those lines, one at a time. Screw and glue the 2x3s in place, and as the first is installed, repeat your straight cut to trim the end to 45°. Do the same to the second.

Glue fine grit sandpaper to the faces of the 45° members.

This jig will work on half strokes (a full push of the board will cut your jig table top in half) to give you perfectly matching miters. To get full strokes, though they're not needed, form a 4" high bridge over the back edge of the jig. If this is done, that bridge needs to be painted bright red and otherwise marked to let you know the saw blade will be coming out its rear. Essentially, the half stroke jig is simpler and safer, and should be used with the blade guard in place.

Butt Joints

Butt joints are the most common joints. They are readily produced, both across and along the grain, on a table saw (Figure 4-27). Rip cuts, cuts that are the table saw's forte, produce along-the-grain butt joints quite well. A decent set of hold-downs and featherboards help maintain cut evenness, while a top quality rip blade gives a good glue line. If a rough rip blade is used, clean up the

Fig. 4-27 Crosscutting is more accurate using a sliding table. Note blade height here, keeping just enough teeth in the material to cut smoothly, but wasting no energy dragging extra teeth through.

glue line on a jointer, though it is often harder to hold a dead straight line on long boards on a jointer than on a table saw.

Long Rip Cuts

For long rip cuts, use a rip blade. Combination blades are fine for shorter rip cuts, combined with cut-off (across the grain) work, but are not suitable for long rips in hard-

Fig. 4-28 Long rip cut in progress. The proper blade is essential to good rip cuts.

woods. Feed fairly fast. Feed is slow on underpowered saws with heavy woods, but ripping as fast as the saw will safely accept the wood is best (Figure 4-28).

In order to feed rapidly, you must use a good rip blade — my preference runs to twenty-four-tooth, carbide-tipped blades with an alternate top bevel grind and a hook of about 20° (similar to many combination blades, except for the hook angle). Crosscut blades have less hook angle, often as little as 7°, while combination blades will have double that number of degrees. Most use an alternate top bevel grind, though. For heavier, rougher work, a blade with eighteen teeth (numbers of teeth relate to 10" blades) and a flat top grind works well.

Fig. 4-29 Spray on oven cleaner. Note gum build up on blade. This is the result of few cuts, but in cedar that was not properly dried.

Fig. 4-30 Wipe oven cleaner with rag, around teeth first where buildup is heaviest.

Fig. 4-31 Finish wiping. Oil lightly afterwards.

Remove Gum from Blade

Regardless of the brand and type of blade used, keep it free of gum. Gum drags, reduces available power, and tends to burn and discolor the wood. There are a number of degumming solutions on the market, most too costly. The simplest and cheapest is to spray the blade with any good oven cleaner (Figure 4-29). Wipe down with a rag or damp sponge, after the recommended soaking time (Figures 4-30, 4-31). Wipe on a light coat of oil or silicone lubricant afterwards, and make the first cut in a foot-long piece of scrap stock to get rid of excess lubricant that might ruin finish or foul a glue line.

Lock Joints

The lock joint, like the finger joint, is an American invention designed to speed up joint making, though the lock joint is a lot harder to make with a table saw than is the finger joint. It requires accuracy of a fine degree, and great care, and in ¾" stock works as follows:

Make the first cut ⅛" deep and ⅛" wide, ⅝" down from the board end (most standard kerf table saw blades offer a ⅛" kerf, but check). Cut a ⅛" x ⅛" rabbet on the same side, top edge. Come back from the other side of the end and cut a ⅛" by ½" deep rabbet, leaving a ⅛" extension. Next, on the second side, cut a ⅜" slot, ¾" deep. The bottom edge is now cut off exactly ⅝" in from the end. Make a second pass to make a ⅛" x ⅛" deep dado above the cut off part, or set your waste removal blade at ⅝" to make that dado cut at the same time you remove the ⅝" of waste.

The resulting joint is tight and strong, and more easily made on a router table. Compared to hand cut dovetail joints, it goes together very quickly with a table saw. To save time, stack pieces for multiple parts in order, and make cuts in sequence on all pieces before moving on to the next setup. You must be dead sure of your setups.

MOLDING HEAD OPERATIONS

Table saw molding heads are popular for making a number of joints on the table saw more rapidly than is possible using just the table saw blade or dado head. Generally, as with tongue-and-groove cutters, each unit is done in a single pass, instead of the two or more passes needed with other setups.

The molding head replaces the saw blade on the saw arbor, and accepts the cutters designed for that brand of molding head. Do not mix brands unless the manuals state you can. For example, most Sears Craftsman molding heads and cutters will work with Vermont-American molding heads and cutters. The molding head doesn't limit you to glue joints of various kinds, as moldings are also possible.

The tongue-and-groove set is a two-piece setup. That is, there are six knives, three for tongues and three for grooves. Boards are set and passed over the grooving knives first, and then set and run over the tongueing knives. The result is a mirror image design.

Molding knives are mounted in trios, equally spaced around the perimeter of the molding head. The slots have setscrews, and many bear on steel balls that seat in the hole for the knife once the setscrew is tightened. Keep a molding head clean. Dirt causes improper seating to mess up cuts, and could do much worse, letting the head fling a knife.

Once knives are in place, install the head on the arbor and place the molding cutter head insert on the saw. If you don't have a cutter head insert, make one. The standard blade insert is too narrow, while the dado head insert is slightly too narrow and too long. The extra length of the slot allows the wood to dip and deform the cut.

Use an auxiliary fence of wood, screwed to the rip fence. Set up the fence cover so you can work on both sides of the rip fence, if needed. Most of us tend to work on the side we're used to, except in extreme conditions.

If you make your own molding cutter insert, use the widest planer knives for the cutout in the insert. Make the insert of the appropriate thickness hardboard or plastic. Place and secure it. Make sure the molding head, with knives installed, is below the surface of the insert, and start the saw. Slowly raise the molding head until it comes through the insert.

The same procedure is used to form the relief cut in the auxiliary fence. The fence needs to be more than 1" thick. For best results, use 1½" to 2" oak or maple. Place the auxiliary fence, and slide the fence over the insert, with the stopped molding head just below the insert slot surface. Raise the running molding head slowly, making a cut the full 1" of width of the cutters.

Molding heads need slow feeds, as do dado heads. The wider the cutters on either, the slower the feed needed. Routers allow a rapid feed even when removing appreciable amounts of material, but table saws work some 15,000 rpm slower than routers, so large amounts of material to be removed need slow feeds of material for an accurate, clean cut.

Avoid Splintering

While on the subject of feeds, consider a feature of wood related to feed speed and cut type and depth. That is the cross-grain decorative or other cut. At the end of a cross-grain cut there will be feathering and splintering of wood, no matter how much care is used.

Among the methods used to get rid of this splintering is the use of a board slightly wider than needed for the finished project. It is then ripped to width. Too, the feed may be slowed down even further, reducing tear out and splintering. When all four sides of a project need cutting, something that often happens with decorative molding and routing, work the cross-grain sides first. The cuts with the grain are used to clean up the splintering (Figures 4-32, 4-33, 4-34, 4-35).

Fig. 4-32 Proper blade angle assures accuracy. This device is costly, but works exceptionally well on 90° and 45° settings.

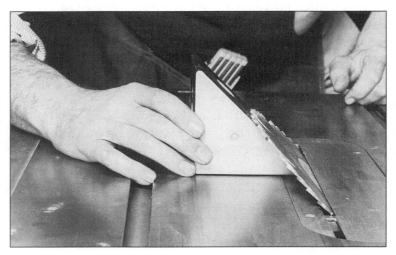

Fig. 4-33 Set at full height, and then reduce just enough to carry one tooth above the material.

Fig. 4-34 Perfect Miter works well with a table saw. Do not forget to install the stops! Otherwise, it's possible to shove the jig into the saw blade, wrecking both.

Fig. 4-35 The resulting miter, on a very quick setup, is near perfect.

OTHER POWER SAWS

The making of joints is not confined to furniture quality construction. Tools less accurate than the table saw are available for use, giving a rougher joint but more speed in construction. There is one other stationary saw that does almost everything the table saw does, with little danger of kickback (the prime drawback of table saws, and a function of blade design and rotation that cannot be totally eliminated), and great ease on jobs the table saw doesn't do well, such as cutting curves.

Using the Band Saw

The band saw has retained its popularity in the home shop, even though getting one of the cheaper models set up can take close to a full year. In addition, esoteric and near nonsensical diatribes on tuning a band saw go as far as using pitch (sound) tuning techniques to get proper blade tension. For those of us with little music education, or a tin ear, there remains the old standby of simply measuring the tautness of the blade. It's almost as accurate, and not nearly as finicky.

Band saws differ from table saws in that the blade does not spin on an arbor, but rotates around two — or in some models three — wheels, and passes through two sets of guides, one held above and another below the table. The blade is a welded loop and may vary in width from ⅛" to as much as 3" on band saws aimed at resawing operations. Resawing is the procedure of ripping a board through its longest dimension, to get two thinner boards.

Band saws can be used to cut lap joints of most styles, though center lap joints need some help from a chisel. Miter joints, all butt joints, and dovetail joints may be cut on a band saw (Figure 4-36).

No special instruction is needed to cut lap, butt, and miter joints, but some extra care is needed with dovetail joints.

Band saws are best used to cut single dovetails, though they may be used, with a chisel, to cut multiples if you wish to do the work that way, or have no other saw. Small multiple dovetails are best cut with a router, or with a back saw and chisel. Larger dovetails are readily cut with both the band saw, and the saber saw. The pin is laid out on the member, with 7° or 9° legs. Feed along the marked lines. Either turn or back out, and come back in on each side of the pin to finish the cuts (Figure 4-37).

Cut the tail section after the pin: mark the tail from the pin, and make a single

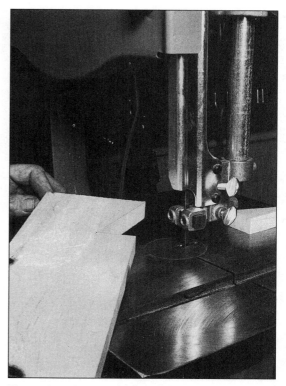

Fig. 4-36 *Cut single dovetails on a bandsaw, if you wish. One side is cut here.*

Fig. 4-37 The exaggerated (for the photograph) single dovetail is cut easily on the bandsaw, with a standard ¼" blade.

cut down one side, curving into the end, or wide, portion that would normally be chiseled out. Carry on across and curve up the other side. You can then make the sharp corner cuts with the band saw or saber saw, or come back and chisel out the excess material. Chiseling is easiest if the piece is fairly long. The saber saw makes it easier to come back and trim out the sharp edges. The blade can be given a tilt to match side spread in degrees, to make a true dovetail.

Test for fit. If it is too tight, loosen a bit with a wood rasp or a surform tool.

Using the Circular Saw

Circular saws are primarily used in construction work to form miter and butt joints. Their accuracy depends on the experience of the operator and the jigs used to assist in the cut. Important too are the quality of the saw, its blade, and blade sharpness (Figure 4-38).

A carbide-tipped blade is usually best. When cutting old wood, or any wood that might have nails embedded, use a standard steel blade designed for cutting flooring. Nails when struck sometimes dislodge carbide tips and send them flying, an experience far better read about than felt.

The blade may be a combination for general use, which cuts the need for either changing blades when making both rip and crosscuts or using two saws. For extra smooth cuts of either type, use the correct blade. For the best miters, use a top grade planer combination blade.

The rip guide, a feature often ignored

on circular saws, is an excellent aid for long rips. This steel guide adjusts to allow the appropriate cut-off of material with the grain, but is not useful for cross-grain cuts. It is easily used. You just slide the fence into the slots on the saw's base, and tighten the screw or screws, once the distance is set. Let the fence guide on the outside edge of the work and make the cut.

Such fences should be used only on material that has an even outside edge. If the edge is uneven, or if the fence will not fit, as it won't on large sheets of plywood, there are other ways to get accurate long cuts. The best is the guide board.

Build a Guide Board

Go to a lumberyard and have a piece of tempered quarter-inch hardboard cut about a foot wide and 5' to 10' long, depending on the lengths of the cuts you need. You may make several of these guides in different lengths. To that tempered hardboard, glue and screw a ½" thick x 4" wide board that is as straight as possible. Fasten along the long side, parallel to and flush with one long edge.

The last step in preparing the jig is simple. Set the jig in place, after measuring, and marking as an offset from the cut line on the board to be cut, the width of your saw base plate to the blade. Clamp it on the material to be cut. Cut both the jig and the material with the circular saw. After this, measure the material, and place the jig so the cut off edge is at the edge of where the cut will be — assuming you always use the same circular saw with this jig. Mark both jig and saw for use together, if

Fig. 4-38 Circular saws may make long freehand rips. Courtesy of Black & Decker.

Fig. 4-39 Circular saws can also bevel well, as this Saw Boss is doing. Courtesy of Porter-Cable.

you have more than a single circular saw.

If you have a hard time measuring the cut width of any circular saw, place the saw on a scrap board of sufficient width and make a practice cut, with the wide side of the base plate resting on the wood keeping the wide edge of the baseplate flush with one edge of the wood. This is usually the motor side. The distance from that mark to the kerf side is the distance you need to place the saw to get an accurate cut.

This measurement also allows you to use nothing more than a straight board to guide the saw, if you wish. Simply clamp the board at the appropriate distance back from the cut line, and guide on the board with the saw.

Make the measurement once and mark it with nail polish or paint on the top of the baseplate for later reference (Figures 4-39, 4-40, 4-41, 4-42, 4-43, 4-44).

Fig. 4-40 Circular saws can free-hand miter as well. Courtesy of Porter-Cable.

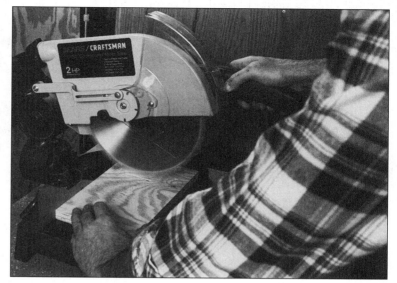

Fig. 4-41 Craftsman's 10" portable compound miter saw works well when freehand miters are not accurate enough.

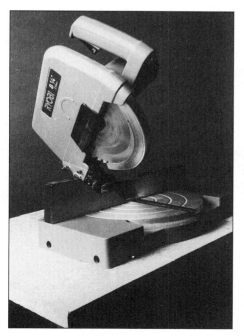

Fig. 4-42 The Ryobi compound miter saw is lighter, but still works exceptionally well.

Fig. 4-43 Jigsaws can cut many joints well. Here, the saw is being used to cut a slot to produce an egg crate style joint.

Fig. 4-44 Dremel's two-speed scroll saw is a slick tool for some joints, including single dove-tails, lap joints, and a few others. Scroll saws will do light band saw work, too.

CHAPTER 5

Making Joints with Hand Tools

ost of our need today are met with machine-made joints, no matter the style. There comes a time, though, when hand-made joints are either more practical or more desirable. At the start of your wood working career, tool limitations may create a need for working with hand tools instead of power tools and costly jigs. If reproductions of older furniture styles are being made, and your desire is to be as accurate to the period as possible, then hand-cut joints are essential.

Use hand tools if setting up the power tools will require more time than hand layout and construction of a number of joints. Thus, if setting up machinery will take more than an hour, use hand tools if the joint can be accurately cut in less than that hour.

Dovetails

The most difficult of all joints to cut by hand is the dovetail. It is also usually the most difficult of machine-cut joints, but as a hand-cut joint, it requires a lot more practice before it starts to shape up as one would wish. When starting this book, I realized it had been years since I had hand-cut a dovetail. My first efforts proved practice was a dire need. Yours will, too, whether you are restarting or starting for the first time. Be a little patient, use cheap wood, and make many practice cuts; your hand-cut dovetails will eventually work well.

Dovetail layout is simplified with the Collet dovetail marking jig, a jig that provides different angles of the pins and tails for hardwood and softwood if desired.

The jig has both 1:8 and 1:6 brass blades in a machined steel frame to allow proper marking of both hardwood and softwood dovetails. Use the 1:6 blade in softwood, with pins and tails of equal size, for maximum strength. For smaller, neater work, the 1:6 blade is used to make finer pins. For really fine work, the 1:8 blade is used to make extremely fine pins that are strong enough for many uses.

Prepare the wood to thickness, width, and length. Make sure the ends to be joined are square, and use a marking gauge to mark a shoulder line around the stock ends of both pieces. Set the shoulder line down the thickness of the wood being used from each end, such as 1" on 1" wood (Figures 5-1, 5-2).

To add strength, mark one piece ⅛" in along the out-

Fig. 5-1 This stock is marked for starting the marking with the Collet dovetail marking gauge.

Fig. 5-2 Front markings are finished here, but need to be carried over the top of the wood.

side edges to add some material to the outside half pins to be formed. Finally, divide the width of the wood by the number of pins desired, after the allowance of the extra ⅛" (¼" total) for each half pin.

Once the number of pins is noted, get the widths, and mark the center points on the board. You now need the tail piece, which is marked first. Move in, on the board, to the first full pin socket, and mark the width. Mark to both sides of the center line. Place the dovetail gauge's center line on the center line mark already made for the pin socket. Lock the steel stops to either side of the center line so that the brass blade moves across the full pin socket width.

Lock the blade in the right position and place the gauge on the first center line (that for the half pin). Mark, preferably with a scriber, along the *left* side of the blade. Repeat the procedure down the pin socket center lines, with the blade locked to the right, and marked to the *left*.

To complete marking, release the blade and mark it against the left side stop. Place the gauge at the first full center line mark, and scribe on the *right* side of the blade. Repeat this across the board; mark the waste to be removed to make sure you do not get confused and cut out the good stock.

Normally, you saw the pin sockets to produce a tail-piece that is used to mark the pin board. You can, instead, use the gauge, setting it at the end of the pin board and making your marks to correspond to center lines as marked. With the blade lined up with the first full pin mark, and in the left position, mark to the right of the blade. Repeat until finished. Relock the blade to the right position, start at the first half pin spot, and mark across, marking to the left of the blade. Mark the waste.

The Collet gauge also serves to mark the end of the board, giving the correct taper to the other side of the board with no extra work.

At this point, the joints are marked and ready for cutting. Use either a tenon saw, dovetail saw, backsaw, or coping saw to make the cuts, and a chisel, coming in halfway from each side, to cut out waste. The coping saw

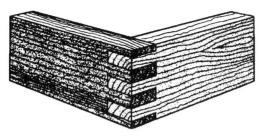

BOX CORNER JOINT

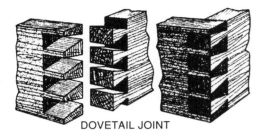

DOVETAIL JOINT

Fig. 5-3 *A box corner joint and a dovetail joint.*

is best used to cut out waste at the bottom, before chiseling. Check fit, and pare waste away until fit is neat (Figure 5-3).

To ease the fit, cut a very light chamfer on the inside edges of the pins and tails. The chamfer will be hidden when the joint is assembled and will provide a pocket for excess glue. Only light coatings of glue are needed on dovetail joints. To prevent glue starvation, especially in hardwoods, coat both surfaces lightly and assemble. If necessary, use a rubber or rawhide mallet to assemble the joint. Check for square, and clamp if you feel a need. The shape of dovetail joints tends to cut out any need for clamping. Put the assembly aside until the glue has set.

The Collet dovetail gauge sounds a lot harder to use than it really is. Two or three practice runs will have you popping off markings in short order, and those same two or three practice runs are needed to gain the handiness with saw and chisel to make the cuts neatly. Remember, with dovetail joinery, to cut to the waste side, leaving the scribe line visible.

Dovetails may also be laid out with a sliding T bevel set to a 10° angle, and a square, using no other layout tools except a pencil or scribe, not even a protractor to determine the angle. Dovetails have one or more pins on the pin member that fit tightly into the tails on the tail member. If a single dovetail is used, there will be two half tails.

For general dovetail layout, set the sliding T bevel to a 10° angle for a correct angle between the vertical line and the sides of a dovetail pin or tail. If you don't have a protractor to set the T bevel, square a line across a board at least a full 6" wide. Mark 6" up the board, and then mark 1" over at the 6" point. This gives a 10° angle with the starting point of the first line, against which the T bevel may be set (Figure 5-4). Measure and mark the shoulder line for the dovetails (usually the thickness of the tail member material plus a fraction of an inch to allow for clean-up: you may disallow the 1/64" or so fraction if you wish). Lay off a half-pin, once pin size is determined, at each edge of the member, and then locate the center lines for your other pins. Lay off the 10° angles to each side of the center lines. Again, carry the angle across the end of the board (Figure 5-5).

Pins are cut with a tenon saw or backsaw, just to the waste sides of the marked lines. Use a scriber to mark lines for greatest accuracy. You'll have to work extra carefully, in good light. Stand the board in a woodworking vise, and cut the pins down. Next, cope saw and chisel out the waste between the cuts. Chisel halfway through from one side, turn the piece and chisel halfway through from the next side. Pare carefully to get down to the line, if necessary.

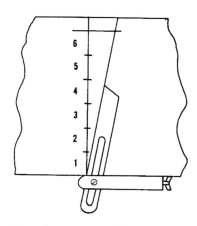

Fig. 5-4 Laying off a 10° angle for dovetail use.

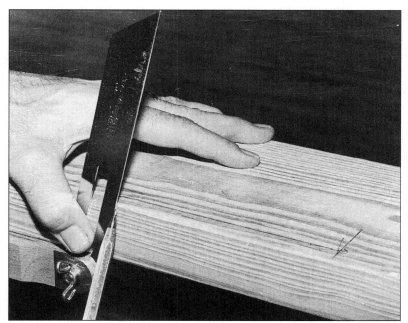

Fig. 5-5 Using a sliding T bevel is the best way of transferring angles.

To mark tails, lay the tail member on the bench and set the pin ends exactly where they will fit. Scribe the outlines of the pins to get the outlines of the tails.

Square a line around the base of the tails at the same thickness as the pin members, and saw and chisel out the waste from between the tail. The procedure is exactly the same as removing the waste between the pins.

Make a trial fit and pare waste away as required to get a good, tight fit.

Box (Finger) Joints

Box joints are basically machine made, and have always been, but they may be made by hand, following the details for through dovetails above. The only difference: do not use the 10° offset. Scribe shoulder lines at the thickness of the material, make the depth of the joint come to those, and make the width suitable for material thickness. Most of us use ½" or ¾" material, for which ¼" and ⅜" finger joints are most suitable.

The joint requires a lot of cutting and chiseling and is less satisfactory than the dovetail but nearly as difficult.

What you get here is a lot of cutting and chiseling for less benefit than dovetails, though the difficulty rises to almost that of dovetail cutting.

Mortise-and-Tenon Joints

Mortise-and-tenon joints are at least partially made by hand most of the time. The final chiseling out of the mortise must often be done by hand to fit the tenon, even if everything else is done by machine.

The procedure changes aren't difficult. The tenon members are supported in a woodworking vise, and are cut with a backsaw, to scribed shoulder lines, to suit the type of mortise-and-tenon joint being made. Clean-up to the shoulder lines is done with an extremely sharp chisel (Figures 5-6, 5-7).

Good old-fashioned chiseling of mortises may be done, and works well, if slowly and sometimes not so

Fig. 5-6 Chiseling a surface mortise for a bench stop.

neatly. I have no objection to using chisels for heavier material, but am clumsy enough to have problems with mortises under ½" or so wide. The work requires more practice than most of us get, especially done in hard to work woods such as oak. Recently, I chiseled a mortise in oak and managed to snap almost all the tip off a top quality Footprint chisel that was quite sharp. At the same time, cutting ⅜" mortises in oak, I have found the wood splitting — a feature of oak is easy splitting.

That may have been a bad chisel, an especially hard piece of oak, or bad technique, but the point is, it was an

Fig. 5-7 The bench stop in the mortise.

incredible inconvenience with the price of oak these days, not to mention the cost of the chisel and wasted work time.

Hand work is fun, but only when it works. Thus, for deeper mortises, I work with a Jet square mortising drill and chisel set, from Trend-Lines, on a drill press. The depth possible is much greater than anything I will need, the cuts are cleaner, and the entire job is simpler.

For hand work, the mortise is marked as for machine work, but has its center drilled with a bit brace and a bit the same width as the planned mortise (Figure 5-8). Drill a hole at one end of the mortise line, another hole at the other end, then drill holes in between overlapping the adjacent holes. Use a mortising chisel to clean the mortise. A bevel edge chisel is used for lighter work, and for paring, positioning the chisel so the bevel side faces the mortise (Figures 5-9, 5-10, 5-11). Use a corner chisel to clean the corners. The entire mortise may be chiseled out without drilling out its center, but that is foolishness, and something I doubt our ancestors would have considered — except as a joke on a new worker.

As with machine work, careful measuring and marking, and careful work with the proper tools, properly sharpened, provide a tight fitting, strong joint.

Fig. 5-8 Bit brace and bits for hand drilling.

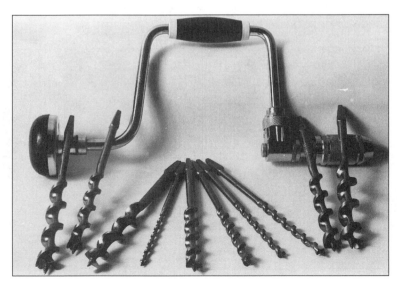

Fig. 5-9 Cheating a bit: drilling extra material out with a drill press. This is to be a blind mortise and tenon, so note the duct tape depth flag on the drill bit.

Fig. 5-10 Heavier chiseling is done with a mortising chisel.

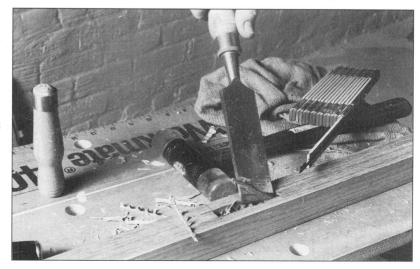

Fig. 5-11 The exact size, or slightly smaller, chisel can be used to cut widths, and trim out what the larger chisel leaves.

Haunched mortise-and-tenon joints are joints that have a cut back section of the tenon to leave a stepped-down design with the mortise cut to fit. This presumably increases strength in some situations, and it usually is cut to avoid making a mortise open on three sides, and thus forming a different kind of joint — a bridle joint.

Probably the most useful mortise-and-tenon joint, no matter how it's cut, is the stub mortise and tenon, where the tenon does not penetrate the stock used for the mortise. This is also called a blind mortise and tenon. There

Fig. 5-12 Note that I've goofed here: the marked narrow end of waste stock should have been cut off first to keep the material marked.

are probably half a hundred variations of mortise-and-tenon joints beyond these, the cutting of which can all be worked out from the basic joints (Figures 5-12, 5-13).

Lap Joints

The lap joint is among the handiest, for a variety of reasons. If nothing else, lap joints are handy braces for two crossing boards. They serve to join boards at ends, and at the end of one board and in the center of the other. Lap joints may join boards of different thicknesses and widths. The instructions for lap joints given here use boards of the same thicknesses, but adjustments are easily made. Decide which side must be flat, and have the less thick of the two boards come flush to that side when joined.

End laps are as easily cut with hand tools as with power tools, and are cut more quickly if only a few joints will be required. Start by making sure the lumber is sized, and the ends are square.

Mark off the lap sizes to half the depth of the boards, marking all sides. Start with lapped boards the same width, so the lap is the width of the boards being joined. Using a backsaw, make a level cut down to the half depth marking at the end point of the joint — as always, leave the scribed line. Repeat the cut several times, leaving strips of whole wood no more than 1" across between cuts. Do the same with the second part of the lap.

Clean out the material between the laps with a chisel, using the widest blade possible. The extra cuts make the chiseling easier, helping to keep the chisel from gouging the wood.

Fig. 5-13 The finished rough tenon.

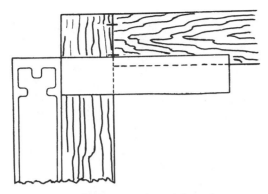

Fig. 5-14 *Indicate on board face the center line(s) for any dowels to be placed.*

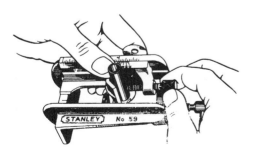

Fig. 5-15 *Select a suitable size dowel for wood size, and select same size drill guide. The guide goes in bevel end up, with the bottom of the guide almost flush with under side of slide.*

Center laps are done in the same manner as end laps, but with two end cuts to mark out the width of the joint.

Mitered joints may be lapped, too.

One of the more interesting hand-cut lap joints is the dovetail half-lap. This joint serves exceptionally well at joining the rails of a cabinet, front to back. It is really a simple joint, with the lapping member having the tail, and the lapped member having the socket. The tail is the depth of the lapped member's width, a point that is marked first. The dovetail is marked out with 10° sides from the outside corners to the marked shoulders. Cut the material out with a backsaw or coping saw, and use the dovetail as a guide for the socket. If the dovetailed piece is not as thick as the lapped member, it remains full depth. If it is as thick, the dovetail is reduced by 50 percent in depth to fit, just as if it were a standard lapped joint.

Edge Joints

Hand cutting edge joints, joints such as the plain butt, the doweled, the tongue and groove, and the spline, is somewhat less practical after one moves past the plain butt and its miter variants. It has been many years since I have seen tongue-and-groove planes for hand use. You can probably find a set if you work the antique shows, and may find a set for sale in ads from time to time, but I know of no manufacturer who makes such tools.

There's really little need, for the use of tongue-and-groove joints is so great that there's not much point to working them by hand. To produce all the flooring, for example, in just a single room would take any of us a

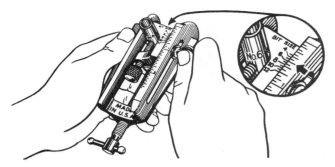

Fig. 5-16 *Now, adjust the slide, aligning the index line for the selected guide, matching numbers to center the hole. As an example, the index line shown, #6, is for a ⅜" guide. If you are setting the guide to center on a 1" (actual) piece of wood, set the guide to ½". For a more usual ¾" piece of wood, set the guide to ⅜".*

good deal of time with hand tools, while machinery can pop the stuff out in short order.

Making doweled joints by hand is no real problem. It's simply a matter of using the same techniques for measuring and siting the dowels as normal, following the instructions on your particular dowel jig, and then using a bit brace and drill bit to drill the holes, instead of using an electric drill or drill press. Dowel centers may be used, if you prefer those to doweling jigs. If it is simpler that way, then fine. Go with it. If not, it takes only seconds to set up an electric drill, and not much longer to set up a drill press for boring dowel holes. Either way, marking dowel holes accurately is far more of a nuisance than is drilling the holes and assembling the work, whether drilling by hand or machine (Figures 5-14, 5-15, 5-16, 5-17, 5-18, 5-19, 5-20, 5-21, 5-22).

Making splined joints by hand is not totally practical, as cutters to let in the slots for the splines are not readily available any more.

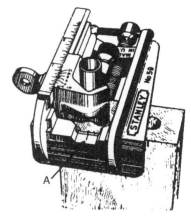

Fig. 5-17 *Place the jig on one of the pieces of material, with the fence next to the face side of the wood. Bring the center line on the guide (A) in line with the mark on the wood. Clamp snugly.*

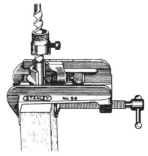

Fig. 5-18 *Place the bit of the correct size into the guide — make sure not to strike the cutting edge of the bit against the guide. Bore to the desired depth — use a depth gauge or a tape flag marker to show depth.*

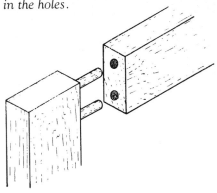

Fig. 5-19 *Check the fit of the dowels in the holes.*

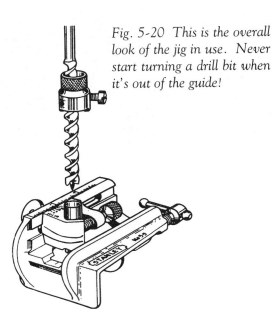

Fig. 5-20 *This is the overall look of the jig in use. Never start turning a drill bit when it's out of the guide!*

Fig. 5-21 *Different dowel patterns can be easily arranged by marking the correct center lines and drilling.*

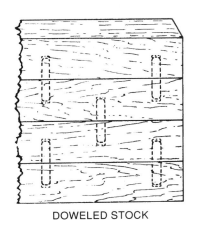

DOWELED STOCK

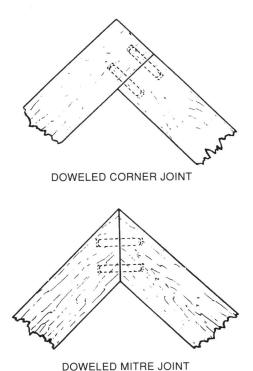

DOWELED CORNER JOINT

DOWELED MITRE JOINT

Fig. 5-22 Doweling jig, ready for use.

Plain old butt joints can be readily cut by hand, using handsaws, and finishing off with planes where super clean edges are required. Miters can also be made by hand, and frequently still are. Top grade miter boxes for hand work cost more than the middle and top grade power miter boxes, and are still made. There are still people who believe in hand sawn mitering. Sandvik's bow saw style miter box (Nobex, but made by Sandvik), is exceptionally handy, as is my heavy Stanley. The Nobex offers a reasonable cut size, with a changeable blade to allow use on different materials, while the Stanley offers a 6" x 30" backsaw with a cut capacity over 10" at 90°, and close to 7" on a 45° cut.

Fig. 5-23 Machine-cut rabbets often need hand help.

Rabbet cuts are handy, and are reasonably easy to make by hand. In most cases, such cuts are made with rabbeting planes, to give the open ledge we call a rabbet (the British call it a rebate) (Figures 5-23, 5-24).

Dado and groove joints are not often done with hand tools these days, though the work is quite possible. For dadoes, a backsaw is used

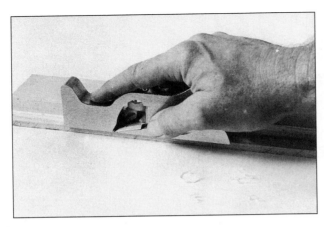

Fig. 5-24 Shoulder planes, such as this model from Woodcraft, clean up here, but can also be used to make the entire rabbet.

Fig. 5-25 Goose neck chisels clean dadoes nicely.

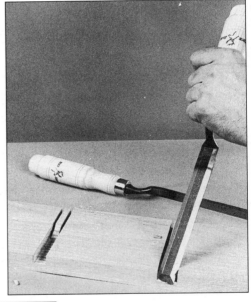

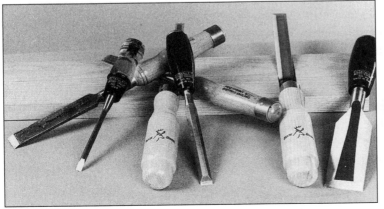

Fig. 5-26 Various chisel styles need to be kept on hand for different jobs, even if very little hand work is done. These are superb for cleaning up, one-time shots, and similar uses, as well as for doing all the work.

to saw, across the grain, to a particular depth, after which the dado's groove is cleared with goose-necked chisels or a shoulder plane, or both (Figure 5-25). Long grooves present other problems, and are usually not worth the effort. Shorter grooves may be cut with a backsaw and chisels, or with a rip saw, which at least has the appropriate, if rough, teeth. Grooves are best started with a circular saw set to the proper depth, after which they may be finished with chisels or planes (Figure 5-26).

Handsaws for regular cuts are readily available, at rational prices, and the new hard point saws, as much as I hate to admit it, do offer some benefits. They come from the maker with a good edge, and hold that edge longer than standard saw blades. The cut they make is straight, and tends to stay that way — if the saw is treated properly. No handsaw should be laid down where other items will be set on the blade; the blade should be lightly oiled after cleaning, and cleaned after each use.

For general and fine work, I still prefer the old-fashioned top grade handsaws such as Stanley's Professional and Nicholson's 300. These saws cost a lot of money, but come in 5½-point rip versions and 8-, 10-, and 12-point crosscut versions, hold a sharpening and set very well, and last for many years. The handles are comfortable in use.

Proper use is simple with any handsaw. Start a cut to the waste side of the cut line, guiding with the first thumb knuckle against the side of the blade until the teeth bite. Guide the saw and cut with the handle at a comfortable angle — this angle varies some for different people, but is generally between 45° and 60°.

Saw care is simple, and includes proper use. Make sure the blade won't strike the ground or other objects under the wood being cut. Make sure there are no nails in the wood. Do not force a saw that kinks in a cut. Back off and clean the gum off the blade, and try again. Often the saw kinks because the board is twisting. Whenever possible, handsaws should be hung up after use. Keep the blade lightly oiled and free of gum and pitch buildup.

Planes must be stored so their irons are not constantly in contact with hard surfaces. Plane irons must be kept lightly lubricated, as must any bare metal on good tools. Chisels may be stored in rolls or racks, or with tip protectors.

CHAPTER 6

Using Plate Joiners

Plate joiners are also known as biscuit joiners, an insult to good biscuits. The "biscuits" or plates are flat and football shaped, and .148" thick, regardless of width. The saw blade in the joiner tools cuts a kerf .156" thick, to provide a loose fit. The plates absorb water from the glue and rapidly swell to over .160".

The plate joiner in portable form hasn't been around as long as many of our other tools, but has been around longer in Europe than in this hemisphere. The plates themselves trace back to 1956, when Steiner Lamello Ltd. (from the German for "thin plate") started making them. Within thirteen years, the company was manufacturing a portable groove-milling machine for the plates.

Lamello started the craze, which first reached the U.S. about a decade ago, but it has stayed relatively dim until recently. When plate joiners first started receiving attention in national magazines, the use of the machines grew greatly.

The growth of Lamello lured other firms into the market, so now Virutex (Spain), and Freud (Italy) have portable models. All of the models from these companies have a number of things in common, including one of the screwiest handle setups anyone has seen on any tool, ever. Add to this an extremely high noise level — machines produced by those three companies *exceed* 100 decibels, a level that is really rough on hearing in short order.

There are two more models of general interest. Porter-Cable came out with a version that solves a couple of problems, and takes a solid bite out of the noise problem. The Porter-Cable 555 is by far the quietest portable unit on the market, because it is driven by a belt, instead of helical gears. The belt drive allows a different shape as well as quieter operation, so that the Porter-Cable model is the easiest to work with because its handle is designed to operate as a handle should.

In addition, the Porter-Cable model is a great deal cheaper than the other versions because of the belt drive.

Delta recently came out with a stationary version of the plate joiner. I have not yet tried it, but Stan Black, president of Trend-Lines, says it is excellent.

All of these models work with three sizes of little plates, to come close to replacing dowels as an aid to wood joinery.

Accuracy

Accuracy of jointing is far better with plates than with dowels. The slot cut to accept the plate allows adjustment along the length of the biscuit, while a dowel pegs you to a point and keeps you there. If you've drilled your dowel holes a fraction of an inch off, your project will be a fraction of an inch off. With plates, it is unlikely you'll be much more than a small fraction off because of the way the joiners are designed, but if you are, you can move things around until the mate is perfect.

Some problems are possible, as my wet shop made particularly obvious. My first load of plates was fine, but I unpacked them and placed them in large plastic bins on my pegboard. A major mistake.

As I have already mentioned, my shop tends to be damp almost eight months of the year. When the biscuits were first opened, it was fall, and things were reasonably dry. A dry winter followed — hot air heat would have kept indoor humidity within rational levels, anyway. A wet spring, an even wetter summer — and the start of a supremely wet fall — followed on the heels of the unpacking, and now there is not a single usable biscuit in the 5,000 or so left from the original unpacking!

The biscuits are designed to absorb water and increase their thickness, by as much as .0015". These did so, which means they won't fit into a .0156" kerf, after

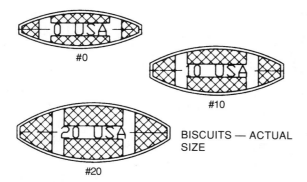

Fig. 6-1 Plates, or biscuits, in the three sizes (#0, #10, #20). Courtesy of Porter-Cable.

#0

#10

BISCUITS — ACTUAL SIZE

#20

starting at a thickness of .0148" (Figure 6-1).

Reduction in size is possible. A microwave oven reduces overall water content and allows biscuits to shrink back close to original size, though the fit, dry, is quite tight afterwards. (You'll also find it doesn't take too long to overheat the plates and stink up the kitchen.)

I will keep the plates in the upper part of our house, to dry out in our forced air heat, after which they'll be squeezed flat in my woodworking vises, and kept in sealed plastic bags, with some form of absorbent material to protect them from future expansion.

Buy biscuits only as needed. Buy them in small packages, and buy only a single package of each type. Do not open any of the packages until you need the biscuits, and then reseal them as soon as possible, and store the packages in a dry area.

I can find no real difference between the no-name brands of biscuits and the more costly brand name types (Lamello, Freud, Porter-Cable, etc.). It makes sense to go with the cheapest available when the only differences are cost and name. The biscuits are of solid beech — stamped to size after being sawn into laths.

PREPARING BISCUIT JOINTS

Biscuits are used to join surfaces, replacing splines or dowels in the process, and usually lead to a neater, quicker job. Gluing is simple. Glue is best dribbled down along

Fig. 6-2 The plate joiner is ready to slot mitered material, which requires a change of position for the front guide plate.

Fig. 6-3 Marking for edge-to-edge joining.

the sides of the slots, after a screwdriver or knife has been used to remove chips from the slots.

As with all joint preparation, the wood must be cut accurately to size, with ends square or mitered as required for the junction to be made. The better the overall preparation, the better the resulting joint, as always.

The more practice cuts you make, within limits, the better you'll find the results of your finish cuts. There is something going round that seems to say written directions and a few markings on a machine will take the place of practice these days. Not so. *Nothing* takes the place of practice.

Cuts are aligned using marks on the joiners. My joiner is the Porter-Cable 555, which operates a little differently (especially with miters) than the others because of its different handle and blade drive system (Figure 6-2).

The joiner has a mark that extends up the front of the baseplate. This is the center point of one's work with the plate joiner. Lining this mark up with one board, and then with a second, means those boards will line up right when cuts are made.

Edge-To-Edge Joining

Mark where you need the plates to add strength to the joint. This is at 8" or wider intervals for edge-to-edge joining (Figure 6-3).

For such joining, work with two boards at a time, no matter how many you're joining. Mark the boards in 2" from their ends, and about 10" apart between those marks. Place the cutting guide for your model, so that the slot is cut halfway down the board's thickness. Cut to the marks on both boards, insert biscuits, and check the joint. If necessary, clean out the grooves.

The final steps are to disassemble, place the glue, insert the biscuits a last time, and clamp the boards together, in alignment. When the glue has set, repeat the process on the next pair of boards, if needed.

Final preparation of the resulting wide board is easy, light sanding. This takes less time than working with most other edge-to-edge glued materials.

Corner Joining

It is in the corners that biscuit joinery really shines. The ease, speed, and results will have you flinging all sorts of doweling jigs in the trash (or at least setting them aside for a yard sale).

Again, align the pieces, but this time the top piece board end is facing you, sitting flush on top of the end of the bottom board and at right angles to that board. Make your marks, starting 2" in from the ends, marking 4" to 6" apart (Figure 6-4).

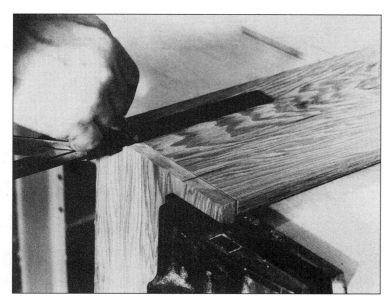

Fig. 6-4 Marking to make a corner joint.

Cut the slots, starting with the face board, then going to the end of the other board. You will be cutting into the end piece with little to brace the joiner when you are using any of the tools except the Porter-Cable 555. This is cured by lining up some bracing material on which to rest the joining tool.

Test fit with dry plates, after cleaning out the grooves if needed. Disassemble, add glue, and reassemble, with clamps.

Center Butt Joining

Joining internal parts with butt joints is easily done, but requires a few more steps. When marking to the center of a ¾" board, mark on the ⅜" line, then always work to the same side of that line. When boards are marked for a particular place in an assembly, key mark them so they return exactly to that place for practice assembly, and for later final, glued assembly. If they're not marked, and the piece is at all complex, pieces will end up in the wrong places.

Cut vertical slots first, after which cut horizontal slots. Marking distances (thus plate insertion distances) are similar to corner joints: come in 2" from each end, then set other biscuits 4" to 6" apart.

If you are having trouble getting all the glued parts together in box assemblies within the ten or fifteen minutes open working time of aliphatic resin wood (yellow) glues, do not just hold things together for ten or fifteen minutes until the glue sets. Instead, get liquid hide glue and take advantage of the far longer open time to get a proper assembly, and to clamp the assembly square.

Miter Joints and Plates

In some materials mitered joints look far better because they provide a finished, all-wood corner appearance, without allowing plies to show with plywood, or to keep grain direction and general appearance more similar in solid woods and pressed boards. As a modified butt joint, a miter joint offers little strength beyond what any

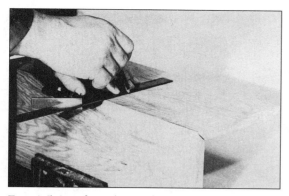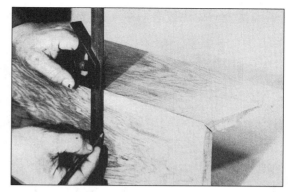

Fig. 6-5 *Marking for miter joints; the most complex of all the very simple marking needs for biscuit, or plate, joinery.*

butt joint does, so it needs some form of support to provide the greatest durability.

Splines and dowels have long been popular for this type of support, with splines taking the prize. A miter joint is set on a table saw so that the blade rips a ⅛" wide kerf the length of the miter. The corresponding board is treated in the same manner. A spline is then fitted into the groove, glue is applied, and the unit is clamped together. Splined joints offer horizontal mobility during assembly so that accurate fit is simple, in that plane. Dowels offer a difficult fit in all planes, especially once the angle changes from 90° to 90° inclusive — that is, 45° + 45°, 30° + 60°, and so on.

Plates offer adjustment along the same plane as do splines, while offering easy and accurate handling of the slotting so that there is less likelihood of a misfit.

Produce a square miter, just as you would normally, in the size required for the job. Bring the two mitered faces together, and mark 2" in from the ends, and after that at 4" intervals. For very large carcasses, place the marks at 6" to 8" intervals (Figure 6-5).

Change the faceplate to the 45° side and make the cuts for the biscuits (Figure 6-6). Make your dry test assembly. This is the time to correct any out-of-square problems created by the biscuits, but if the cuts are ac-

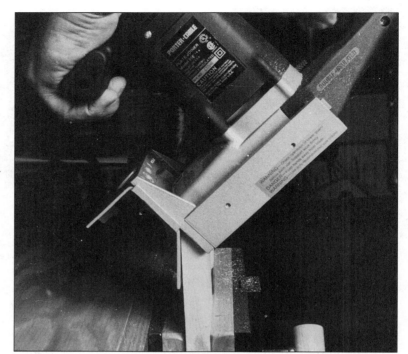

Fig. 6-6 *Slotting in progress.*

curately made, and the miter was done properly, there should be no problems (Figures 6-7, 6-8, 6-9). Disassemble, glue, and reassemble. If you have trouble clamping such assemblies, try a band clamp, like the Vermont-American model shown, which is the easiest I have seen to use, and which requires no extra tools. Check for square as you clamp.

Plate joiners offer a number of options to increase strength with thicker stock. It is quite possible to offset cut marks even in fairly thin (¾") stock so that dual biscuits may be used, one above the other. It is seldom essential, and not a good idea. Internal pressures can cause an imprint of the plates on the surfaces.

In thicker stock, extra strength could well be needed. If we assume stock thickness of 1½", the offset will normally be ¾". By making it ⅜", as if for ¾" stock, we can make one cut, on a marked line, then turn the board over and make a second cut on the same vertical marked line.

Fig. 6-7 Miter ready for test assembly.

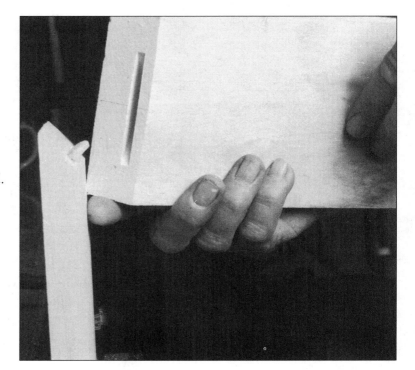

Fig. 6-8 Test assembly started.

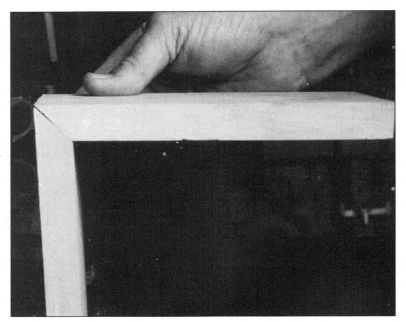

Fig. 6-9 Test completed. Glue up can proceed if this is the extent of slot making.

That produces two slots for plates, and allows use of two biscuits per area, with a normal amount of stock outside the biscuits, as well as between the biscuits.

You may wonder if the cost of one of these joiners is worth it. That is a decision only the person in charge of the wallet can make, but for anyone doing a lot of butt, dowel, and spline joinery, the answer is yes. Shop around. Prices vary by quite a lot.

Buy biscuits in 1,000-unit bags instead of 5,000 unless you have assured dry storage, or an immediate need for a great many biscuits. But you will be amazed how fast you go through the things after you get used to your plate joiner (Figure 6-10).

Fig. 6-10 With the front guide removed, it is possible to note blade tip of the plate joiner.

CHAPTER 7

Mechanical Fasteners

Mechanical fasteners include a great many items, not just a range from screws to nails and back again, with some variation on those. If the variation isn't in the screws or nails, then it is in the devices to be used with them, such as knockdown fasteners, or what are called mending plates.

Mending plates are flat metal plates that are screwed over a joint to brace that joint. They may be used to mend, but are just as often used on a new joint to brace. They come in Ts, Ls, Hs, and a variety of other shapes that prove useful from time to time. Newer types are introduced all the time, as are nails, screws, and knockdown fasteners. Keep checking your local stores and wood-working mail order catalogs to see what is available in any mechanical fastener line.

Wood Screws

There is a wide world of wood screws and screw-type fasteners that may be used to fasten wood or objects to wood. A number are especially popular, and have become so for good reason. Today, changes in the field are more frequent than in the past as new tools bring out new driver shapes, which require different head designs on the screws.

The essential changes today lean toward screw types meant to be power driven. With the array of cordless power drivers now on the market, cam-out resisting screws are needed more than ever. Cam-out is the twisting out of a slot or other style head of the driver tip as more power is applied, with resulting damage to the material being used. It is more likely to happen with slotted screws.

Without this feature, torque twists the driver tip out of the slot, and the tip goes on to mangle the screw head or to mar the surface around it. Cam-out is far more common with power drivers, because of higher speeds and torque ratings. There is a strong emphasis on Phillips head and square insert screws for use with both hand and power drivers, while the old standard, the slotted head, stands on the side of the road and watches the others pass it by. This creates no real problem, because any of the new screws is adapted to use with hand drivers as well as power drivers. The cost differential is also very small, where it exists at all (Figure 7-1).

Screw Types

Generally, screws may be classified as wood, lag, or metal screws. Otherwise, if threads are involved, it's nuts and bolts (bolts do not have tapered screw shanks, and accept nuts). Wood screws may have round, flat, or oval

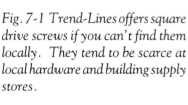

Fig. 7-1 Trend-Lines offers square drive screws if you can't find them locally. They tend to be scarce at local hardware and building supply stores.

heads, while metal screws offer pan, flat, and round heads, and a variety of others of little or no immediate interest. Lag screws have square or hexagonal heads, with coarser screw threads than wood screws (Figure 7-2). Lag screws are offered in longer sizes, up to 16" in length and 1" in diameter. Most hardware stores carry sizes up to about 8" or 10" and ½" shank diameter. Larger than that and you'll wait a few days for the special order to come in. Wood screws run up to 6" in length, and #24 in size. The diameter is expressed in numbers, as Table 7-1 shows, and these numbers relate only indirectly to the actual fractional inch sizes. Commonly, you'll find wood screws up to about 4" or 5" in length, and #16 in size in most hardware stores, in a number of materials. Other sizes, and extremely small sizes such as ¼" in #0, #1, #2, or #3, have to be ordered.

Wood screws are usually made of mild steel, coated or uncoated (zinc or galvanizing), solid brass, or stainless steel. Coated and solid brass varieties are useful where

Fig. 7-2 Types of screws.

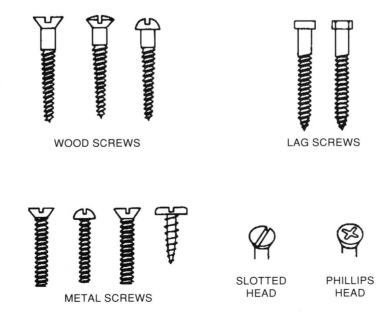

WOOD SCREWS   LAG SCREWS

METAL SCREWS   SLOTTED HEAD   PHILLIPS HEAD

TABLE 7-1
SCREW SIZES AND DIMENSIONS

▼ LENGTH (IN.)	SIZE NUMBERS											
	0	1	2	3	4	5	6	7	8	9	10	11
¼	X	X	X	X								
⅜	X	X	X	X	X	X	X	X	X	X		
½			X	X	X	X	X	X	X	X	X	X
⅝		X	X	X	X	X	X	X	X	X	X	X
¾			X	X	X	X	X	X	X	X	X	X
⅞			X	X	X	X	X	X	X	X	X	X
1				X	X	X	X	X	X	X	X	X
1¼					X	X	X	X	X	X	X	X
1½					X	X	X	X	X	X	X	X
1¾						X	X	X	X	X	X	X
2							X	X	X	X	X	X
2¼							X	X	X	X	X	X
2½						X	X	X	X	X	X	X
2¾							X	X	X	X	X	X
3							X	X	X	X	X	X
3½									X	X	X	X
4									X	X	X	X
4½												
5												
6												
▶ THREADS PER INCH	32	28	26	24	22	20	18	16	15	14	13	12
▶ DIAMETER OF SCREW (IN.)	.060	.073	.086	.099	.112	.125	.138	.151	.164	.177	.190	.203

Table 7-1 (cont.)
Screw Sizes and Dimensions

▼ LENGTH (IN.)	SIZE NUMBERS									
	12	13	14	15	16	17	18	20	22	24
¼										
⅜										
½	X									
⅝	X		X							
¾	X		X		X					
⅞	X		X		X					
1	X		X		X		X	X		
1¼	X		X		X		X	X		X
1½	X		X		X		X	X		X
1¾	X		X		X		X	X		X
2	X		X		X		X	X		X
2¼	X		X		X		X	X		X
2½	X		X		X		X	X		X
2¾	X		X		X		X	X		X
3	X		X		X		X	X		X
3½	X		X		X		X	X		X
4	X		X		X		X	X		X
4½	X		X		X		X	X		X
5	X		X		X		X	X		X
6			X		X		X	X		X
▶ THREADS PER INCH	11		10		9		8	8		7
▶ DIAMETER OF SCREW (IN.)	.216		.242		.268		.294	.320		.372

corrosion resistance is essential. A plated zinc or galvanized coating is normally used. Stainless steel screws are useful when corrosion problems are extreme, such as in or around salt water, and in high acid environments. Steel screws are also used where strength greater than that of brass screws is required. Brass screws are the weakest of the three generally available wood screws, but are decorative and corrode very slowly. Stainless steel corrodes hardly at all, but is the most costly material. Mild steel, even zinc plated, is the cheapest material.

Wood screws come in sizes which vary from tiny ¼" items to screws up to 6" long. For screws to 1" in length, the step increase in length is ⅛", while screws from 1" to 3" long increase in length by ¼" at a step. Screws from 3" to 6" jump in ½" increments. Shaft sizes vary with the number used to specify such size. Shaft size numbers are arbitrary, but rise with increasing size, and, as Table 7-1 shows, there are no #13, #15, #17 or #22 sizes.

Power-drive screws, both fine and coarse threaded, come in sizes somewhat different than those used for commonly available wood screws. Drive screws tend to be available in longer lengths — up to 3", for example, in light shanks, say a #6, and seldom larger than a #9 — where common wood screws would have thicker shanks for the length (usually at least a #10, most often a #12, and frequently a #14). The reason is that the thinner shanks create less driving resistance.

Power-drive screws are a relatively new product, and have either a Phillips or a square drive head. They are meant only for power driving, and are very useful for installing decking, wallboard, and wall paneling of certain types, and for general light construction duties.

Most, with the exception of some specific types, offer poor shear strength, so even a large number of the screws do not serve as a replacement for a single lag screw of the same length when installing a ledger board to support a deck or other heavy structure.

Screws are more expensive than nails, but have some benefits for the extra cost and work. Holding strength is a

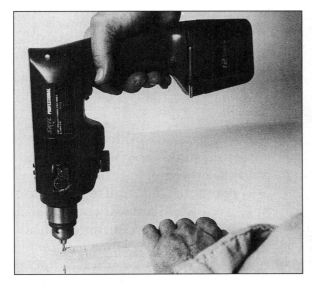

Fig. 7-3 Drilling a pilot hole in flat stock.

lot greater. Disassembly is a great deal easier, without destroying the item being disassembled.

Screws are more work to install. For most screws, a pilot hole will do, but for flat head wood screws, countersinking is essential (Figure 7-3). Flat head wood screws are used for decorative purposes, where screws are not meant to show, and are often counterbored, and the resulting hole covered with a plug that may be flat to the surface, or domed (Figure 7-4).

When drilling pilot holes, drill the hole at least one size less than the screw threads of the screw in hardwood, and two sizes in softwood, and make the hole a half to two-thirds as deep as the screw will sink. On occasion, in hardwood, you'll need to drill the pilot hole as deep as the threads will reach (Figure 7-5).

Other Screw Fasteners

Machine screws are used with nuts, and often washers, to join wood to metal or wood to many other materials, including wood. Machine screws come in different materials, and are used with different kinds of knockdown fasteners to produce an easily assembled and dis-

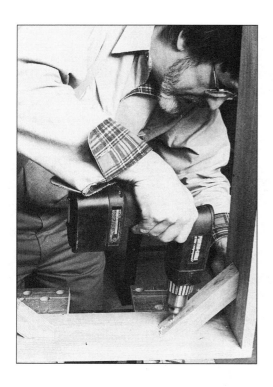

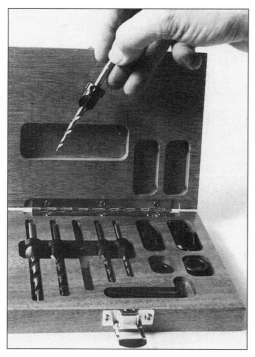

Fig. 7-4 Pilot holes are essential on angled surfaces, for even if the screw or nail doesn't split the surface, it may skitter off a number of times before going in at the wrong angle and position.

assembled project. Among the popular holders for machine screws are T nuts and brass screw inserts.

T nuts are a type of nut that fits into a drilled hole in one wood surface. They are set in place and then tapped down so the teeth in the upper ring grip. The screw is then run into the T nut center as if it were an ordinary nut, allowing sides and other assemblies to be mated, with an ease of knockdown built in for future needs, when a project might need to be moved in flat condition, or parts of it might need to be changed (Figure 7-6).

Brass screw inserts work in a manner similar to T nuts as far as holding power goes, but are inserted by screwing them into a hole drilled to size. The inserts have coarse male threads on the outside, and finer female threads on the inside. The brass inserts are screwed into the holes. Their tops are slotted to accept a standard flat blade screwdriver tip. The insert is turned down until the top is flush, then a brass or steel machine screw is driven into its internal threads.

Fig. 7-5 Tapered drill bits offer screws the best grip, so are great for pilot hole drilling. This set comes with countersinks and stops. The sets are available from mail order sources such as Woodcraft, Trend-Lines, and The Woodworker's Store.

Fig. 7-6 The nuts serve many purposes in making joints, and even in inventing joints.

For insert fasteners, you can use knurled screws. Knurled brass screws are decorative, but also ease disassembly because no screwdriver is needed to install or remove them. They serve well as pivot points where projects need such points. Use a slightly longer knurled screw to fit into the brass insert, and place washers on both sides. The knurled screws are available in ½" and 1" lengths, in 8-32 and ¼-20 threads (Figure 7-7).

Similar fasteners are used for coarser jobs. These are called joint sets instead of inserts and are meant for use with special steel connector bolts. An Allen wrench is used to install them, and the inserts are a ¼-20 thread that will accept 2" or 2¾" bolts.

Fig. 7-7 Brass inserts serve the same purpose as T nuts, but somewhat more elegantly. The machine screws and knurled screws, and the inserts, are available from Trend-Lines, if you can't find them locally.

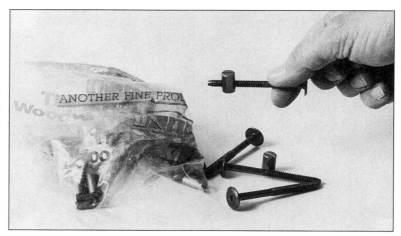

Fig. 7-8 Cross dowel sets are one type of knockdown joint fitting available from The Woodworker's Store.

For further knockdown construction, check cross dowel sets. These are units that consist of ¼" x 2" or 2¾" bolts, and the cross dowel nuts that fit on the bolts (Figure 7-8). The nuts are set into ½" deep holes drilled into pilot holes (a ¹³⁄₃₂" bit is needed), and an Allen wrench is used to drive the headed bolts. Another version of this exists for different knockdown uses. Instead of a cross dowel, there is a cap nut that threads in to set two flat boards face to face, rather than boards at right angles to one another.

Fig. 7-9 Tite-Joints are another type available from The Woodworker's Store.

For general knockdown construction, I doubt there is a better roundup of the varied available types of fasteners than you'll find in The Woodworker's Store catalog. Some of these are so specialized there's little point in covering them, but there are enough to suit almost any job where you may want to disassemble it later (Figure 7-9). Both Trend-Lines and Woodcraft also carry moderate ranges of knockdown and other fasteners.

Other Threaded Fasteners

Lag screws are far more important in heavy wood joinery than most people think, serving to hold roofs onto house sides, decks on, porches on, and serving all sorts of heavy anchoring jobs against masonry projects, when used in conjunction with lead screw anchors.

Lag screws — correctly termed but almost never called lag bolts, wood screw type — are heavier than common wood screws, and have either a square or hexagonal head. The coarser threads extend down to a gimlet or cone point, with the threads covering slightly more than half the screw length. A wrench is usually used to drive the lag screw, and the head should come to rest on a washer, not on wood. Lag screws serve where ordinary wood screws are too short or too light, and where driven spikes won't provide enough holding power (Figures 7-10, 7-11, 7-12, 7-13).

Carriage bolts give us another method of fastening things together, and are generally intended for use in wood or metal, where the holes are drilled all the way through the pieces to be connected. There are three basic types of carriage bolts, each type based on neck style. The square or common neck bolt, finned neck, and ribbed neck all have a rounded over head (Figure 7-14). The necks on carriage bolts are intended, when used in wood, to be

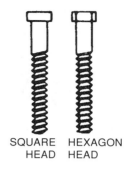

SQUARE HEXAGON
HEAD HEAD

Fig. 7-10 Lag screws.

Fig. 7-11 Lag screws and washers. These are fender washers, extra wide types.

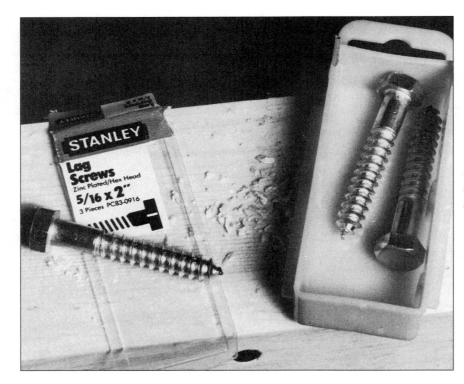

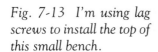

Fig. 7-12 Heavy cross section, hexagonal head lag screws.

Fig. 7-13 I'm using lag screws to install the top of this small bench.

drawn up tight into the wood to prevent the head from turning. The shank fits through an exact size head. In fact, the carriage bolt usually needs to be tapped through its holes with a hammer. A slight counterbore is often used to fit the neck of the carriage bolt, though in some softwoods, smaller carriage bolts will often draw down flat with no counterbore. They may also split the wood, so counterboring is a good idea. The nut screws on, with a washer used under the nut, regardless of whether it covers wood, metal, or plastic.

Finned and ribbed neck carriage bolts may have to be special ordered, though they are pretty widely available in some sizes. Overall, carriage bolts come in lengths from ¾" up to 20", and in diameters from ³⁄₁₆" to ¾" (Table 7-2).

TABLE 7-2
CARRIAGE BOLTS

▼ LENGTHS (INCHES)	DIAMETERS (INCHES)			
	³⁄₁₆, ¼ ⁵⁄₁₆, ⅜	⁷⁄₁₆, ½	⁹⁄₁₆, ⅝	¾
▶ ¾ ----------------------	X	----------------------	----------------------	----------------------
▶ 1 ----------------------	X	X	----------------------	----------------------
▶ 1¼ ----------------------	X	X	X	----------------------
▶ 1½, 2, 2½, ETC., 9½, 10 TO 20 ----------------------	X	X	X	X

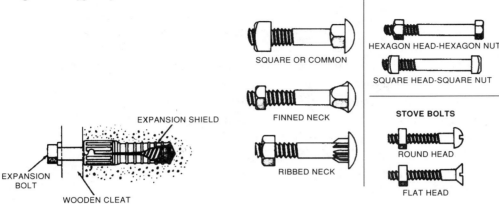

Fig 7-14 *Types of bolts.*

Machine Bolts

Machine bolts can be used in place of carriage bolts in wood if you have access to both sides of the work so both head and nut may be turned with wrenches. Machine bolts come in more sizes than do carriage bolts, ranging from ¼" to 1¼" in diameter up to 39" in length, and may have either hexagonal or square heads. Square nuts are best used with square heads, and hexagonal nuts with hexagonal heads. The bolts will usually be supplied that way. Common flat washers are used under both head and nut. Machine bolts generally have closer tolerances than do carriage bolts, for their primary use is for metal to metal joinery. General installation, including tapping the bolt through its drilled hole, is similar to installation of carriage bolts, except that wrenches are needed on both the head and the nut.

Nails

Like the butt joint, the simple nail provides security for most wood joinery today, though it does seem that no one really classes carpentry as wood joining. It is. Carpenters used to be called carpenters and joiners, a phrase that has slipped away with our modern oversimplification of language — and skills, in many cases.

The fact remains that the nail is probably still the most useful mechanical fastener around, as well as being the easiest to use and the least strong in many applications (Table 7-3). It makes up for a lack of strength with a low cost and easy, fast installation. In many applications today, the power-driven screw is replacing nails. Using power drivers, screws can be run in almost as quickly as nails or staples, and fewer are needed to provide greater strength, and thus at least partially making up for the higher cost of the fasteners themselves.

Iron nails go back as far as the Roman occupation of Great Britain and quite probably well back from there. Roman nails were forged, creating differences in shape caused by manufacture, but in general were similar to today's wire nail.

Cut nails took over from wrought or forged nails with most of the change coming in the late 1700s and early 1800s. The early cut nails had handmade heads, but by about 1830 the entire nail was cut by machine. (You can, if you wish, buy machine cut nails today from Tremont Nail Co., P. O. Box 111, Wareham, Massachusetts 02571.) Cut nails provide an antique appearance and greater holding strength than common wire nails. Their holding strength is as much as 70 percent greater, according to cut nail makers.

Wire nails, though, are our modern fasteners, and supply the majority of fasteners in stick building as well as in the remodeling industry and some other parts of furniture, frame, and other woodworking industries. Wire nails began as cheaper substitutes for cut nails, but early nails didn't look much like what we find today. Those early wire nails were not too well made, with bulbous heads that were usually off center on the shank.

At the outset, wire nails were primarily used in building of small boxes, not construction, but by 1888, the size range had increased, with thirteen sizes ranging from 2d nails to 60d available.

TABLE 7-3
SIZE, TYPE, AND USE OF NAILS

SIZE	LGTH. (IN.)	DIAM. (IN.)	REMARKS	WHERE USED
2d	1	.072	SMALL HEAD	FINISH WORK, SHOP WORK.
2d	1	.072	LARGE FLATHEAD	SMALL TIMBER, WOOD SHINGLES, LATHES.
3d	1¼	.08	SMALL HEAD	FINISH WORK, SHOP WORK.
3d	1¼	.08	LARGE FLATHEAD	SMALL TIMBER, WOOD SHINGLES, LATHES.
4d	1½	.098	SMALL HEAD	FINISH WORK, SHOP WORK.
4d	1½	.098	LARGE FLATHEAD	SMALL TIMBER, LATHES, SHOP WORK.
5d	1¾	.098	SMALL HEAD	FINISH WORK, SHOP WORK.
5d	1¾	.098	LARGE FLATHEAD	SMALL TIMBER, LATHES, SHOP WORK.
6d	2	.113	SMALL HEAD	FINISH WORK, CASING, STOPS, ETC., SHOP WORK.
6d	2	.113	LARGE FLATHEAD	SMALL TIMBER, SIDING, SHEATHING, ETC., SHOP WORK.
7d	2¼	.113	SMALL HEAD	CASING, BASE, CEILING, STOPS, ETC.
7d	2¼	.113	LARGE FLATHEAD	SHEATHING, SIDING, SUBFLOORING, LIGHT FRAMING.
8d	2½	.131	SMALL HEAD	CASING, BASE, CEILING, WAINSCOT, ETC., SHOP WORK.
8d	2½	.131	LARGE FLATHEAD	SHEATHING, SIDING, SUBFLOORING, LIGHT FRAMING, SHOP WORK.
8d	1¼	.131	EXTRA-LARGE FLATHEAD	ROLL ROOFING, COMPOSITION SHINGLES.
9d	2¾	.131	SMALL HEAD	CASING, BASE, CEILING, ETC.
9d	2¾	.131	LARGE FLATHEAD	SHEATHING, SIDING, SUBFLOORING, FRAMING, SHOP WORK.
10d	3	.148	SMALL HEAD	CASING, BASE, CEILING, ETC., SHOP WORK.
10d	3	.148	LARGE FLATHEAD	SHEATHING, SIDING, SUBFLOORING, FRAMING, SHOP WORK.
12d	3¼	.148	LARGE FLATHEAD	SHEATHING, SUBFLOORING, FRAMING.
16d	3½	.162	LARGE FLATHEAD	FRAMING, BRIDGES, ETC.
20d	4	.192	LARGE FLATHEAD	FRAMING, BRIDGES, ETC.
30d	4½	.207	LARGE FLATHEAD	HEAVY FRAMING, BRIDGES, ETC.
40d	5	.225	LARGE FLATHEAD	HEAVY FRAMING, BRIDGES, ETC.
50d	5½	.244	LARGE FLATHEAD	EXTRA-HEAVY FRAMING, BRIDGES, ETC.
60d	6	.262	LARGE FLATHEAD	EXTRA-HEAVY FRAMING, BRIDGES, ETC.

Nails are sized by the penny, abbreviated "d," a method once used by manufacturers to determine how many cents 100 nails would cost. Given inflation over the past century, the actual costs changed rapidly, but the sizing system has hung on, with sizes still ranging from 2d to 60d, though many manufacturers prefer now to sell by length and weight (Figure 7-15).

Under 2d or 1", nails are classed as brads, and over 60d (6") they're classed as spikes, though a few makers carry sizes to 80d (8") as nails. Given the effort required, anything over 30d is best classed as a spike when driven. Specialty nails abound.

Common nails are used for general purpose nailing from framing work on through some types of flooring installation. Shank styles differ. Greater holding power is found with deformed shanks such as ring and screw, while coatings are available and the nails may often be hardened. Common nails also come in aluminum as well as galvanized, for outdoor uses.

Box nails are similar to common nails, but have slightly larger head sizes to shank diameters.

Post and truss nails are similar to common nails, but have a spiral or screw shank. They are designed to keep a post-and-beam structure from giving in to wracking stresses. Such nails are now called deck nails, and do a lot to prevent popping of decking nails in such construction.

Wood siding nails are slim-shanked models designed to help prevent splitting of wood that needs to be nailed near its edges or ends. It still makes sense to blunt the nails and, when within 1" of ends or edges, drill pilot holes. Lengths run from 2" to 3½".

Finishing nails are slim, nearly headless nails in sizes ranging from the 1" brad finishing nail on up to at least 16d (3½") and beyond. They may be found in clean mild steel, galvanized, and hot dipped galvanized, and are used because their small heads are easily set (with a nail set) below a board's surface, where a touch of putty eliminates any nailed look.

Finishing nails provide good, clean joints in miters

and similar cuts, if the material being joined is accurately cut. Common nails are useful in framing of houses and other projects, and siding nails do the obvious. Deck nails help in large outdoor projects where wood is affixed to wood.

Corrugated nails, or fasteners, are often placed under the "other fastener" category since they do not even look like nails. Corrugated nails look like small bits of down-

Fig. 7-15 Sizes of Nails

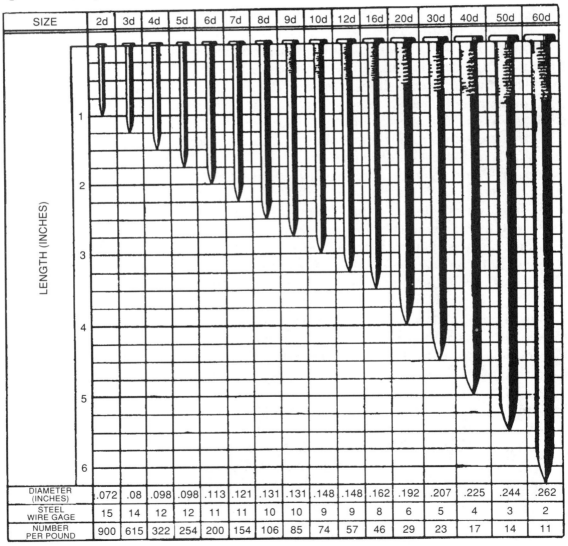

SIZE	2d	3d	4d	5d	6d	7d	8d	9d	10d	12d	16d	20d	30d	40d	50d	60d
DIAMETER (INCHES)	.072	.08	.098	.098	.113	.121	.131	.131	.148	.148	.162	.192	.207	.225	.244	.262
STEEL WIRE GAGE	15	14	12	12	11	11	10	10	9	9	8	6	5	4	3	2
NUMBER PER POUND	900	615	322	254	200	154	106	85	74	57	46	29	23	17	14	11

scaled corrugated roofing that has been clipped off at a sharp angle. They are driven into butt and miter joints to brace the joint, and may be set below the surface of the wood and covered. Generally, they are made of 18 and 22 gauge metal, in varying widths, from ⅝" to 1⅛". Length may vary from ⅝" to ¾" (Figures 7-16, 7-17).

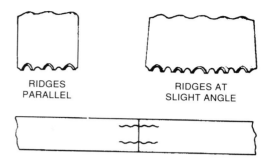

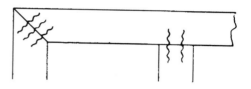

Fig. 7-16 Corrugated fasteners (nails).

THREADING WOOD

There is a variety of wood-fastening devices that may be of interest from time to time, though some are totally esoteric and others are marginally useful.

I've found that a wood-threading device, a tap and die set for wood, is handy for certain projects, and simply used. My set is from Woodcraft, and I only have one size, though it is available in three. The threadbox (corresponding to the machinist's die) is of

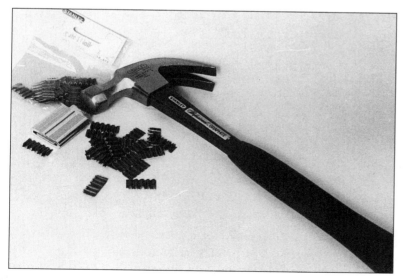

Fig. 7-17 Corrugated nails and nail set. The nail set simplifies getting this fastener to drive evenly, and reduces wood splitting.

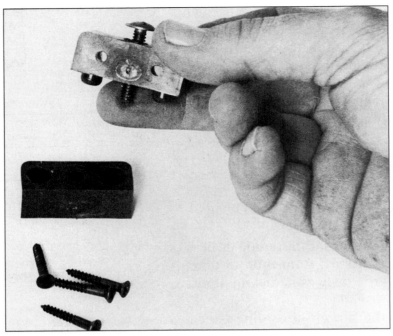

Fig. 7-18 *This knockdown fastener is another one of the many available from The Woodworker's Store. It installs quickly and easily to form right-angled joints for cabinet or desk sides.*

Fig. 7-19 *Stanley Hardware produces both of these brace/fastener styles, with the larger in galvanized steel and the smaller in solid brass. Many more shapes and styles are available.*

maple, with an aluminum insert. The steel tap is cut to quite close tolerances. To use the device, you hold a dowel in a vise, with the dowel size the same as the size of the thread required. Turn the tap box down on the dowel until the required thread depth is reached. Next, take the block into which the thread has to turn, and drill an undersized hole, after which the tap is run down in the hole. The two pieces will screw tightly together.

There is a light production wood-threading device from Beall. I've seen this in action, and it appears efficient and fast, but I have never used it. It uses a router and the Beall jigs to rapidly create wood threads, but is of use only if you're doing a great deal of threading (Figures 7-18, 7-19, 7-20).

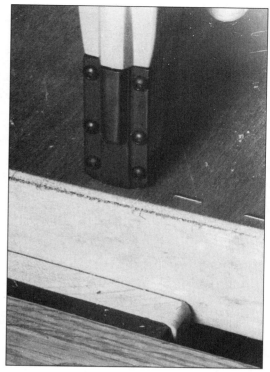

Fig. 7-20 Stapling is a method of fastening wood often overlooked by the amateur. Even though about half of modern furniture seems to be stapled together, it's a process I don't like.

CHAPTER 8

GLUING AND CLAMPING

For most of woodworking history, hide glue was the epitome of wood-joining strength, even with its multiple problems. It also has multiple advantage, as we shall see. Since then, things have changed considerably, though hide glue continues to be the adhesive of choice for certain jobs, and could serve the rest of us with greater frequency than it does.

The primary use of wood glues is to hold joints together. There are other applications, but the major use is making strong wood joints in furniture and other assemblies. Proper gluing will relieve some of the internal stresses of wood, particularly in laminating flat boards and reducing warping, cupping, and other forms of distortion. Too, adhesives allow us to use many parts of a board that are otherwise bits and pieces of scrap.

WOOD SELECTION

Successful joints do not always start with tightly cut joints, well mated and coated lightly with an appropriate adhesive, firmly clamped, and permitted to dry. You must make sure that the woods being joined will successfully mate. Too great differences in the moisture content of woods will create problems, as may similar differences in wood structure. As an example, teak, with its high silicone and oil contents, does not bond well with other woods and is even difficult to bond to itself.

The best wide area glue-ups (especially as sizable laminates or flat boards) result when the same species of wood is used. As examples, all the boards are pine or fir or oak or cherry. If part is cherry and part is pine, difficulties arise. Avoid the problem.

Whenever possible, use the same species, but if different species are desirable, use solid sections of those species where glue is not necessary, if possible. Door construction provides a good example, with the inner panel floating loose in the stiles and rails. The glued inner panel may be of any species, while the stiles and rails may be of a different species. Because the two sections are not solidly joined, there are no joint problems.

For the same reason, use plainsawed boards with plainsawed boards, and quartersawed boards with quartersawed. Otherwise, the differences in grain directions will create distortion problems.

Allow boards to "temper." That means leaving them at least twenty-four hours in the environment in which they will be glued. Two to three days is better.

Selecting a Woodworking Glue

Regardless of claims, there aren't as many true woodworking glues as one might expect. Even fewer are of interest to the general woodworker producing strong joints.

Contact cements are used for general woodworking, and will be briefly covered here. The joint produced using them is not a true woodworking joint; it is a laminate of dissimilar materials, and properties in the cement allow it to last even though expansion and contraction rates differ markedly.

Woodworking adhesives are broken down into animal or hide glues and synthetics. Animal glues are far older, and are far less used today because synthetics offer certain properties they do not. Similarly, there are some properties offered by animal glues that synthetics don't offer, or don't offer as completely.

No one really knows when the first hides, hoofs, and bones of animals were boiled to produce glue. These glues are byproducts of the meat and tanning industries, and are readily found in dry, granular forms or as ready-to-use liquids. The most common liquid hide glue today is Franklin's. The glue is nontoxic. Its slow drying time gives a long assembly time for complex projects, allowing adjustments and changes that can't be made with faster setting glues.

HIDE GLUES

Hide glues are thicker than white and yellow glues, resist solvents (other than water) well, and give a pale tan glue line. They sand well without gumming. The lack of gumming is important, because glues and adhesives that gum heavily will clog and ruin sandpaper quickly. Water resistance is poor for hide glues.

Dry hide glue is available. I suggest that you avoid it. It must be mixed with water and heated, and then be held at 140 to 150°. The process starts with soaking the granules in cold water. The granules soften after several hours, and the excess water is poured off. The glue is then heated, and the temperature maintained while the glue is stirred until it is smooth and free of lumps.

Hot hide glue is applied with a stiff brush. The joint is clamped while the glue is still hot. Wood being glued should be warmed to at least 70° in cold shops. Neither the wood nor the glue should be overheated, as that ruins strength.

Using dry flakes for making hot hide glue is a complex procedure that requires good advance planning and a costly heating bucket. As a result, the use of dry glue is not common among today's woodworkers, though it is far from a lost art.

Synthetic Adhesives

Most of the adhesives used for woodworking today are synthetics, formulated specifically for different applications in the woodworking fields — and some originating in other fields. Most are types of resin glues that gather strength by chemical reaction, or curing. Curing is dependent on the temperature of the glue. The strength of the cure and the speed of setting are increased by raising the glue temperature. Heating the glue to over 120° isn't a good idea, as maximum cure temperature is about 110° for yellow (aliphatic) resin glue, while others do best at 75 to 80°.

Polyvinyl Acetate Resins

White glues (polyvinyl acetate resins) come ready to use, in squeeze bottles on up to gallon and larger jugs. There are many brands, and many of them are acceptable for general woodworking purposes.

White glues don't use a chemical reaction type of cure. The water in the glue moves into the wood and the air, thus the resin gels. On unstressed joints, you can usually release clamping pressure inside of forty-five minutes. Leaving clamping pressure on for several hours is better. Stressed joints need at least a six-hour set before clamps are released (Figure 8-1).

White glues are not always dead white. Some are dyed close to yellow and some tan to appear more like aliphatic resin glues. Aliphatic resin glues are more heat and water resistant.

White glues have poor sanding qualities (Figure 8-2). The reason is that the glue softens because of the heat generated by sanding, so it gums up the sandpaper. The same characteristic causes a loss of glue strength at 100° and up. Water resistance is low enough so that a high humidity basement can create separation problems, as with hide glue joints.

Set is fast, limiting assembly time to ten or fifteen

Fig. 8-1 Resin glues bond edges well, as they do other areas.

minutes. Pressure application must be fast, so preassembly of projects is essential. Fit all clamps within about a half turn before disassembling, applying adhesive, and reassembling for clamping.

The glue line is close to transparent once the white glue has dried. In too cold a shop (under 70° usually, though I've successfully used this type of glue at 65° with some frequency), the glue will appear a chalky white, and the joint line may be weakened.

White glue gives with the day-to-day movement of the wood, a process known as cold flow, so should not be used on a highly stressed joint, structural laminates, for example, where a great deal of bending pressure might be applied and released with great frequency. Cold flow allows joints to move naturally without creating cracked glue lines and weakening the joint.

Liquid Yellow Glues (Aliphatic Resin)

Aliphatic resin glues were designed as improvements over the polyvinyl resin glues, and do provide some worthwhile changes.

Heat resistance is higher, which makes it easier to sand while also improving strength at 100° and up. These glues set well at temperatures up to 110°, which means

Fig. 8-2 Sanding qualities vary, from glue to glue.

they can be used on hot summer days. Assembly is easier at lower temperatures because raising the glue temperature speeds the set rate and reduces open time. The glue line is a translucent pale tan or amber color.

Yellow glues are less likely to run and drip than white glues because their basic consistency is heavier. This makes neater gluing jobs. Their greater moisture resistance means you can assemble projects for use in damp basements within reason.

Set is faster than for white glues. This can be a problem if you are gluing complex projects. Switch to hide glue when project assembly will take more than five minutes.

Total cure is at least twenty-four hours. Use water to clean up before the glue sets (Figure 8-3).

Waterproof and Water-Resistant Glues

A number of glues are available for use where moisture is a problem. Two of the most useful are plastic resin and resorcinol resin glues.

While plastic resin adhesives are highly water resistant, only the resorcinols earn a waterproof rating. Where there is some trade-off possible, it's more economical to

Fig. 8-3 *Interior glues all clean up with a damp cloth.*

use the plastic resins, since resorcinols cost three or four times as much as the plastic resins.

Resorcinol resin glues are dark-red liquids (the resins) to which a catalytic powder is added before use. Resorcinols have a reasonable working life, depending on formulation, after mixing, from about a quarter of an hour to two hours. It is best to work with the longer life, so check the labels first.

Before starting, get well set up. Make sure wood moisture content is below about 12 percent, joints are tight and precise fitting, and heavy-duty clamps are on hand.

Brush resorcinol on or spread it with a spatula. Tongue depressors, available at any drugstore, make great glue

Fig. 8-4 *Resorcinol glues require mixing of powder and liquid resin.*

spreaders, as do ice cream sticks from the grocery store, and toothpicks for small projects (Figure 8-4).

If dense hardwoods are glued, watch out for glue starvation — the lack of glue in tight fitting joint areas. Lightly coat both surfaces with glue, and leave the joint open for the maximum time before clamping. Increasing shop temperatures helps, especially in colder shops.

The project must be clamped as quickly as possible. Clamping pressure must be high, about 200 pounds per square inch. The pressure must also be more uniform. That means more clamps.

The glue line is ugly, a dark red or reddish brown.

Urea (Plastic) Resin Adhesives

Plastic resin adhesives are dry powders that are mixed with water just before use. The resin is urea formaldehyde, and is a highly water-resistant adhesive, best on wood with a moisture content of no more than 12 percent. Best use and cure temperature is 70°.

Plastic resin glues are superb for producing joints in projects that must withstand long-term dampness. Some of these glues do very well in true exterior applications. They make good general-purpose glues because they work easily in all situations, with the exception of high-density woods such as maple and oak. Precise fit of joints is essential, as plastic resin glues are not good gap fillers. The best gap filler, other than epoxies, is hide glue.

Setting the glue is affected by temperature, so complex assembly jobs may be left to cooler weather or done in an air-conditioned area. Otherwise, working life ranges from one to five hours. Clamp pressure must be in place for at least nine hours, preferably twelve. Clamps may be taken off when the glue squeezed out is hard.

Clamp pressure should be moderate. The joint line appearance is good, a light tan color. Gumming is not a problem as the resin resists heat well.

Epoxy Adhesives

Epoxies weren't used in the woodworking shop for a great many years. Recently, formulations and techniques have been developed to aid in using these adhesives. Like resorcinols, epoxies come in two parts, with a liquid hardener (or catalyst) added to a liquid resin. There is no powder. Curing is by chemical reaction. Heat is given off as the reaction takes place.

Mix only as much as you will use. The material is costly, and waste is expensive.

Epoxies are the adhesives of choice when bonding teak and similar high-oil woods. Epoxies can be formulated to suit just about any bonding need, in moderate temperature applications, if you precisely follow package directions.

Epoxy doesn't shrink, so is a good gap filler. Some epoxies are available as putties, thus the largest gap is readily filled, though precise joint fits are still better for long project life.

Set time is an important factor in wood adhesives. One of the reasons epoxies missed early popularity was that virtually all of them are fast-set types, setting in under five minutes. Such speed is superb for many jobs, including some small woodworking projects and repairs (Figure 8-5), but is a horror story for larger projects. There

Fig. 8-5 Quick set epoxies in tubes are very handy for small jobs, and should be kept around the shop for general repairs as well.

is no way to get a large project together and squared up in the available time. Thus, epoxies were useful for repairs, and useless for most other purposes.

With slower set formulations, epoxies became more useful. Epoxies aren't useful, or financially possible, for general woodworking jobs such as bonding strips of wood to form a butcher block, or bonding wood for a table top, or coating a lot of joints for strength. The cost is extreme, and the need simply isn't there (Figure 8-6).

Epoxies are quite toxic, too, which limits their uses in some shops, mine included. Epoxies are messy, though that problem is easily solved. Wear thin plastic gloves, now available in packs of 100, to avoid the hand mess. Clean up quickly with acetone for other messiness, keeping the gloves on. Make sure all mixing containers and sticks are disposable. When you can't keep from making a mess, make the mess itself as easily disposable as possible.

Some epoxies are fine for one job, while others are best suited for others. Keep a quantity of quick set epoxy, in tubes, on hand. Epoxy in tubes is easier to use than epoxy in other dispensers, such as twin nozzle setups. With the latter, eventually you'll lose the leftover portion as some of the catalyst manages to slip into the resin tube, or some of the mix manages to clog the tips.

For filling gaps, waterproofing, and, in my case, as-

Fig. 8-6 Slow set epoxy formulations are less costly per ounce (but still expensive), and give much more time for assembling larger projects. Woodcraft offers a good selection of slow set epoxies, including the G-2 I'm getting ready to use here.

sembling outdoor signs, you'll find the problems of using epoxies are worth facing because of the excellent results.

If you're working with dense woods or with exotics such as rosewood, teak, and ebony, nothing surpasses epoxy as an adhesive. Clamping pressure is light, working time is adjustable, depending on the system, to as much as ninety minutes, gap filling is superb, strength is incredible, and the resulting glue line is either clear or an amber color, depending on the brand used. Some brands won't stain wood, others will. Check. Some may be used with a variety of catalysts, with the catalyst providing either a slow or fast cure. Fillers can be found, as can pigments.

Epoxies are not for general use butt joints, nor general use box joints, but for special woodworking uses, they've really come into their own.

Cyanoacrylate Adhesives

Super, or crazy, glues are better than they used to be for woodworking purposes.

That's not really saying much because they were almost useless as a wood adhesive as tap water. Any product that gains a reputation that exceeds performance and has serious shortcomings (lack of water resistance) hidden among more dramatic, though less important, shortcomings (the ability, presumably, to bond flesh with extreme speed), tends to deserve knocks, and gets them.

Super glues suffered from that, and from the fact that the original formulations were not blended to work with porous materials. Failures with wood were constant.

These adhesives are extremely costly. Current versions work better, but the prices haven't dropped, though they are a little cheaper per unit when bought in large volume.

For model building and any work with miniatures, the super glues can prove invaluable. No clamping is needed, so deformation and breakage from weighty clamps is avoided by coating the surfaces and holding everything together for sixty seconds or so, using only the fingers as clamps.

Fig. 8-7 *Newer, larger pack cyanoacrylates are better on wood than any past types, but still serve best for lightweight work and for reducing or eliminating clamping needs.*

If the job is not delicate, there is no need for cyanocrylates.

If materials being joined are not porous, very little of the adhesive is required. Too much causes a failure to set, or a poor bond. If materials are porous, you need to add more of the glue, effectively sealing the surfaces to a non-porous condition where the final light coat of cyanocrylate bonds (Figure 8-7).

Special uses, faster bonding, and clean-up to speed bonding are provided by various kinds of accelerators. Accelerators speed the cure and bond of special types of cyanoacrylates.

Hot Melt Adhesives

Hot melt adhesives are available in stick and sheet forms, often with the sheet forms supplied as the backing to edging of different kinds. I work with oak plywoods, so oak edging in 250' rolls, with adhesive already in place, is a boon. This joins easily with an electric flatiron. Do not use the good iron; buy a new one for the house and use the old one in the shop.

Working with oak edging is my favorite use for the material, though I have used gun-melted glue to stick shingles on birdhouses and dollhouses, among other things. The gunned hot melt is fine for setting up single cuts of double materials, without nail holes. It works well at holding small items in place for routing, too, and is easily peeled off later. Depending on the formulation, hot melts set in from a couple to thirty seconds.

I prefer it for temporary joints as the overall joint strength, regardless of manufacturers' claims, is far lower than that normally required of good woodworking joints.

Hot melt glue does not sand for beans. It gums up in a hurry, as friction from the sandpaper creates heat. Slice off any residue for easy cleanup.

Hand pressure is all that is required to get a bond, but you'll find things work a lot better if you're in a hot shop, upwards of 85°. This warms the wood so the hot glue doesn't set too rapidly, creating a messed up bond.

Contact Cements

If you've ever built kitchen cabinets, you have probably worked with contact cement. Contact cement usually doesn't go into the cabinets (newer so-called European styles, with laminates over wood substrates, do use contact cement over almost the entire cabinet), but into the counter top where plastic laminates such as Formica are glued to wood substrates. The wood substrates may be plywood, or one of the formed boards. The latter are usually best because, over the years, plywood grains show through the plastic laminates.

Contact cements come in two basic types. One uses a water base, or water solvent, while the other uses a nonflammable solvent base. UGL's Safe Grip uses 111 Trichloroethane as the solvent. Some professional types still use flammable solvents — avoid these at all costs!

Vapors are harmful with the newer solvents, so make sure you work with proper ventilation. Some water-based solvents are pretty rough in the fume field. A few may not be safe around an open flame. Check before using any of them. Avoid flammable contact cements.

Contact cements give a quick bond that allows cleaning up and trimming of the final project right away. Using them is simple. Coat both surfaces with the cement, using a brush or a roller. Let the surfaces become dry to the touch. Place the laminate on the substrate (Figure 8-8).

You must make sure the laminate is positioned correctly over the wood. A slip sheet of kraft paper or waxed paper may be used, covering the entire surface. Leave enough to grip outside the two pieces being joined, bring the top piece down, align the two pieces, and slowly start

Fig. 8-8 Contact cements work with plastic laminates and wood substrates of many types.

slipping the paper out. Once the paper is out 3", roll or tap over the cleared area to assure a bond. Pull the paper the remainder of the way out, being careful of alignment. Roll or tap to assure the bond, and you're done.

You might also use 1" square wood stickers at intervals. Placing stickers across the full width of the base material works well at 1' intervals. The coated laminate is laid on the stickers and carefully aligned with the base. Remove, first, the center sticker so the laminate touches the base material coating. Tap or roll, and then remove the remaining stickers, tapping and rolling as you go.

Choosing Glues

The selection of the appropriate glue is important, but so are the application of the glue and the clamping of the parts. Equally important is that you are working with a tight fitting joint so there are no problems with gap filling or joints weakened by wide expanses of nothing but glue. We want our joints to last, even if we're impervious to the embarrassment of a project's leg sliding loose as Great Aunt Matilda parks her 200+ pounds on it.

Make your glue selection based on the qualities you need most.

If assembly is complicated and time-consuming, and moisture is not a problem, choose hide glue over synthetics. If moisture is a moderate problem, choose a urea (plastic) resin glue instead of yellow glue.

For general uses, liquid yellow glues, and polyvinyl acetate (white resin) glues are fine, with white glue for longer assembly times, and yellow glue for slightly better moisture resistance, better sanding, and thicker spreading qualities that provide better gap filling.

For great water resistance, select either epoxy or urea resin — urea resin first, unless the epoxies fill some other specific need such as great unsupported strength or filling a gap. Epoxies cost too much for general use.

For total waterproofing, select resorcinol resins. These are expensive and difficult to apply properly since they require very precise fitting joints, and they leave an ugly reddish glue line, but they are totally impervious to water.

The type of glue chosen influences the method of application, though most can be applied with a brush, stick, or roller (Figure 8-9). Joint surfaces first must be checked. If the joint surface is designed to be a tight fit, it should be. Clean off all dust, oil, old glue, loosened and torn grain, and chips. Any machining that must be done should be done as close as possible to the time of gluing and assembly.

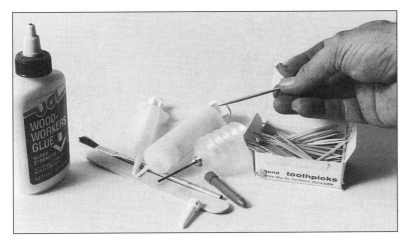

Fig. 8-9 Varied glue applicators are needed in every shop. The injectors, whether squeeze bottle or push down handle, work well for moderate amounts, but need cleaning after each use. Tongue depressors work fine on small to medium amounts, and can be found in any drugstore. Small brushes can be bought by mail order, while toothpicks are probably in your kitchen.

A test assembly is a good idea, because once the glue is added, correcting mistakes is, at best, very messy. If mistakes get in and glue sets, mistakes remain.

Before you apply the glue, decide whether the unit can be assembled within the time required for a specific type of adhesive. If a glue has a ten-minute open time, you must have the assembly completed before that time passes. The thicker the glue you spread, generally the longer the open assembly time. If wood is extremely porous or extremely dry, open assembly time decreases (Figure 8-10).

Fig. 8-10 Glue will work better if applied before assembly.

If you can't make the test assembly within the allotted time, change either the conditions of gluing or the type of adhesive used so enough time to complete and clamp the assembly is available.

Mix, where required, all adhesives according to the maker's directions, and as accurately as possible. Spread evenly over the surfaces to be joined (Figures 8-11, 8-12, 8-13).

Clamping and Clamping Pressure

You clamp a glued joint for three reasons. The wood surfaces must be brought into direct and close contact

Fig. 8-11 I spread this glue with a finger because I forgot to keep a brush nearby. A finger makes a great spreader for interior woodworking glues, assuming you keep a damp cloth handy.

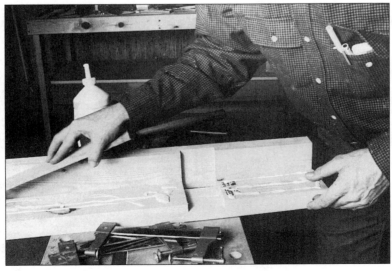

Fig. 8-12 Cedar shim stock works as a spreader, too.

Fig. 8-13 *Fast spreading is needed, and glue must lightly coat both surfaces.*

with the glue, the glue must become smooth, flowing to a thin, continuous film, and the joint must be held steady until the glue dries.

Clamping pressure varies with glue type, but generally coincides with glue thickness and wood type. The heavier the glue, the more clamping pressure. You want a thin, smooth glue line, not a joint that is squeezed dry, which occurs when too much pressure is used.

Most woodworking glues on softwoods fall in the intermediate thickness range, requiring clamp pressure of 100 to 150 pounds per square inch. Some dense hardwoods may require pressures up to 300 p.s.i. Softwoods do not require much clamping pressure and will not stand

Fig. 8-14 *Bessey's K body (or bar) clamp is for heavy duty use, and works very nicely indeed.*

it. At the very least, some deformation will occur with clamping pressures up near 300 p.s.i. on softwoods.

A clamp may apply force ranging up to a ton. This pressure is divided by the overall size of the clamped area to determine the pounds per square inch being applied. Home woodworkers may occasionally apply too much pressure using hand-tightened clamps. This happens most on large surfaces with closely arrayed bar clamps, but it's unlikely.

Resorcinol and urea resins require a great deal of pressure, while epoxy needs little pressure.

Avoid excessive pressure in favor of even pressure over the entire area. It is better to get even glue squeeze out over the entire joint than to rack the project up as tight as possible.

Some woodworkers say one thing, some say another when the numbers of clamps to be used comes into question. I suggest clamps every 8" to 10" when possible, with some light clamps used as often as every 4" to 6". In no case do I feel secure with clamps spaced out more than 16" apart.

Even contact cements require about 50 p.s.i. for a proper bond. That's the reason you need a heavy roller

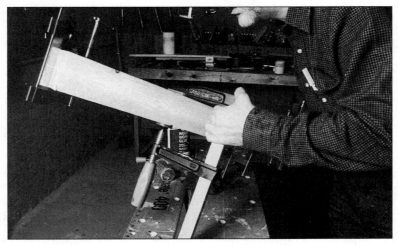

Fig. 8-15 Sandvik's heavy duty bar clamps are for just as heavy duty uses, but work on shorter jobs, as the maximum available opening is not as great as with the Bessey. Nice working clamps.

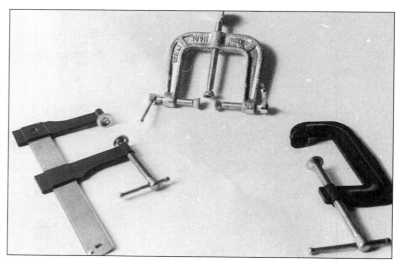

Fig. 8-16 Handscrews work well on irregular surfaces such as this, with one flat, one pointed surface.

Fig. 8-17 A C clamp is on the right, while Stanley's short, lightweight bar clamp is to the left. Both are low cost, and I keep several dozen on hand for different jobs. The edge clamp is in the center; fewer of these are needed by most of us, but they are invaluable for edge gluing strips to plywood, etc.

or a 2x4 and a hammer. If you don't have a laminate roller, use a rolling pin with a sheet of moderately hard, heavy cloth wrapped around it. It can be used without the cloth, if a few gouges and some dirt are not a problem. If you use a roller as thick as a rolling pin, bear down hard to get the correct amount of pressure per square inch.

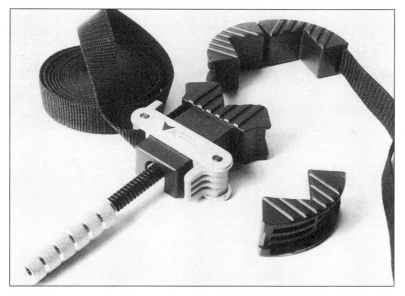

Fig. 8-18 Vermont-American's band clamp offers an internal plastic clamping shaper, or the simple band. Bands and similar clamps are great for such work as chair leg assemblies.

Clamps and Tools

Many styles of woodworking clamps are available, and few of them have changed much in recent decades. Of the new items, nothing is really earthshaking. Woodworking clamps fall into one of these categories: bar clamps (Figures 8-14, 8-15), hand screws (Figure 8-16), C clamps (Figure 8-17), and band clamps (Figures 8-18, 8-19).

Among these you find pipe clamps, devices for

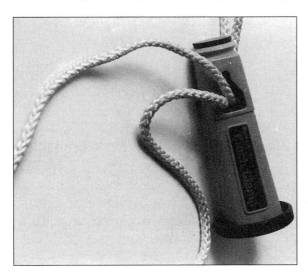

Fig. 8-19 Black & Decker's Cinch (bard) Clamp may be difficult to find. If you can't find one, use 5⁄16" nylon rope to go around the item being clamped, and twist it tight with a stick. Jam the stick back into the rope to keep it from spinning loose. Such clamps are ideal for repairs on chair legs and similar spots where lightweight, irregular holding clamping pressure is needed.

Fig. 8-20 K body Bessey clamps do a fine job of aligning cradle sides, quickly and easily. Those spindles made this assembly touchy though only the end joints were glued.

Fig. 8-21 The cradle was finished fairly quickly after the side assemblies were done.

clamping picture frames, miter clamps, and a slew of others. The largest number of variations fall in the bar clamp range, which makes sense since those are used for case framing, large assemblies of multiple piece tops, and sides. American Clamping Corporation's catalog shows nine variants of bar clamps, without considering sizes and accessory assemblies. ACC has assemblies that will accept almost any case size, that will set edge pieces, and for many uses. Adding sizes to the nine bar clamps produces over six dozen variants.

Sizes for the bar clamps vary with the weight of the clamp. The heavier the clamp, the longer it can be, so normal weight clamps reach about 48", while light duty units stop at 32". Super light duty notched cam and tongue bar clamps stop at

Fig. 8-22 These edge clamps are being readied for use.

24". The heavy-duty Bessey K body clamps ACC offers extend to 98". These are costly, silky working clamps superb for heavy work that also do very well in medium work (Figures 8-20, 8-21).

Edge clamp accessories are offered in single and dual models, and allow a great many variants that standard edge clamps, such as the Jorgensons I have, do not. The edge clamp accessory fits onto a standard bar clamp bar, allowing a great deal of positioning leeway. My edge clamps are fine heavy-duty units, but do not offer as much leeway. They were not designed to do so and do their particular job very well (Figures 8-22, 8-23).

Fig. 8-23 Edge clamp in use. Times like this, water solubility of glues is a real help.

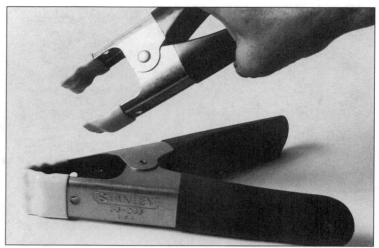

Fig. 8-24 Small and large Stanley spring clamps with padded jaws. Do not buy unpadded jaw styles for woodworking uses.

Band clamps vary. One of mine is about 1" wide and 10' long. Another one is 2" wide and 14' long. The first one is from Stanley Tools, and the second from Vermont-American. Both are excellent tools. Band clamps are useful for clamping odd shapes, such as octagon frames, chair leg assemblies, and picture frames. Some are made with angle sections that slip onto the bands to fit different angled corners.

Short, lightweight bar clamps are a help in some areas that are usually thought of as C clamp territory, such as

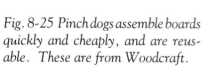

Fig. 8-25 Pinch dogs assemble boards quickly and cheaply, and are reusable. These are from Woodcraft.

gluing two sheets of plywood. It takes just as many clamps, but there is more depth of throat, and somewhat thicker overall clamping possibilities.

Hand screws in this country bear the Jorgenson brand name. I bought mine through Shopsmith some years ago. Tips for preparing hand screws help preserve the screws, and the finish on some projects.

Hand screw bodies are wood, thus susceptible to gluing and other problems should you be messy. To keep this glue buildup from becoming a problem, coat the outer jaw with paste wax and don't bother to polish it off. This means virtually nothing will stick, including glue, but over the years this does create enough slickness to make use of the hand screw tips difficult. Small objects tend to slide out. Get 2" or 3" wide masking tape and cover the front of the hand screws where glue is most likely to spread accidentally. Strip the tape off when it becomes loaded, and replace.

Hand screws work well to clamp non-parallel surfaces, and do not creep as some clamps will. They are available in more sizes than are listed in catalogs. Prodding local suppliers may get you a chance at the larger sizes you need and cannot find otherwise.

This is far from an exhaustive look at clamps, missing a lot of types including corner clamps, useful for mitered work such as picture frames, spring clamps, and the wide variety of levered hold-down clamps. Hold-fast clamping devices for workbenches are also missed; a couple of types are shown in photographs (Figures 8-24, 8-25).

INDEX